The Green Gold of China

绿色中国

John MacKinnon and Wang Haibin
EU-CHINA BIODIVERSITY PROGRAMME
Translated by Lu Hefen and Wang Zanxin
马敬能 王海滨 著
中国－欧盟生物多样性项目
吕和芬 王赞信 译

Foreword

for Creen Gold of China
EU-China Biodiversity Programme publication

Conserving biodiversity is a fundamental part of development. China's exceptionally high biodiversity makes it even more relevant. Through this beautiful book, we can spread this important message.

The EU-China Biodiversity Conservation Programme(ECBP) is the EU's largest biodiversity programme worldwide. It underlines the EU's recognition of biodiversity's importance and of the great economic value of the products and services it provides.

The book highlights the great beauty and value of Chinese biodiversity. It also tells the story of threats posed to biodiversity by China's rapid development, the measures being taken to protect biodiversity and the specific activities being undertaken by the EU-China Biodiversity Programme. I offer my best wishes to all those working on the important task of conserving China's green gold. I hope the expertise that the EU brings to China in this cooperation will be fruitful and leave a lasting legacy.

Serge Abou,Ambassador
European Commission Delegation to China and Mongolia

Serge Abou,Ambassador
European Commission Delegation to China and Mongolia

赛日·安博大使
欧洲委员会驻华代表团团长

前言

绿色中国
中国—欧盟生物多样性

　　生物多样性保护是社会发展的一个重要组成部分。中国的生物多样性极其丰富,这使得中国的生物多样保护在社会发展中的地位尤为重要。希望通过这本精美的书, 这一重要信息能够得到广泛的传播。

　　中国—欧盟生物多样性项目（ECBP）是欧盟在全球范围内资助的最大规模的生物多样性保护项目。该项目充分表明了欧盟对生物多样性的重要性及其提供的产品和服务功能的重要经济价值的认可。

　　本书不仅突出了中国丰富多彩的生物多样性,还强调了其不可替代的价值 同时也讲述了中国快速发展给生物多样性带来的威胁,中国对生物多样性保护所采取的措施,还有中国—欧盟生物多样性项目在生物多样性保护方面开展的具体活动。在此,我衷心地祝愿那些工作在保护中国生物多样性第一线的"绿色卫士"。我希望通过这一项目,来自欧盟的专业知识能够使中国的生物多样性保护获得显著的成效,使后代永久受益。

赛日·安博大使
欧洲委员会驻华代表团团长

目录
Table of Contents

目录
Table of Contents

PART ONE
The Richness of China's Biodiversity

Introduction

Like the light shimmering off the scales of a dragon, China reflects many colours. To the yellow banners of its imperial past and the red banners of modern China is added the green banner of environment. Of course China was by nature green—a vast expanse of forested mountains and verdant grasslands. We took green for granted. It was our natural inheritance. But 5 millenia of human development and endeavour have changed the nature of the land forever. If we are to retain any of the original green or if we are to make green once more the degraded deserts and denuded hills, that will be done by human effort. Our new challenge—Green China.

China is blessed with a diversity of ecosystems, habitats and living species second to none. Whether we use this wealth wisely or whether we let it slip through our fingers is in our own hands. This 'greenness' of China is very precious. It is gold. Its value is counted in economic benefits, ecological services, human health and living standards as well as spiritual well-being.

As a consequence of the vast size, varying nature and complex geological history of China, a very wide range of habitat types and very large numbers of species are represented—from the high peaks of the Himalayas to the depths of the Turpan basin; from the deserts, steppes and meadows to a range of forest types, wetlands, coasts and coral reefs.

Approximately 35,000 species of higher plants belonging to 353 families and 3,184 genera (of which 190 are endemic) occur here. China contains the richest temperate regions in the world and ranks globally as one of the richest countries in terms of overall plant diversity. The country is also recognised as one of the world's ten 'mega-diversity' countries on account of its rich vertebrate and other zoological wealth.

China has about 11.5% of the world mammal species, 13.5 % of the world bird species, 6.7 % of world reptile species and 6.8% of the world amphibian species. The Chinese fauna is also characterized by high levels of endemism or creatures found nowhere else in the world.

Turnstone has a chisel-shaped beak, well adapted for prizing shells off rocks and rolling over small stones to search for hiding crustaceans underneath.

翻石鹬的凿形嘴，可从石块上撬取贝类，能翻滚小石块搜寻藏匿其下的甲壳类动物为食。

第一章
中国丰富多采的生物多样性

导　言

　　正如龙鳞片上闪亮的微光一样，中国反映了多种色彩。在旧日中华帝国时代皇家尊崇的黄色旗帜，以及现代共和国高高飘扬的五星红旗之外，另一面绿色的环保旗帜也已在中华大地树立起来。当然，远古的中国也是绿色的—广袤的森林和翠绿的草地。中国大地当然是绿色的，它是我们的自然遗产。然而，五千年的人类发展已经改变了大地的本色。若要想保留一些原始的绿色，让退化的沙漠和光秃的丘陵重现昔日的绿色，就需要我们的努力。现在我们正面临着一个新的挑战：绿化中国。

　　中国拥有世界上最丰富的生态系统、栖息地和物种多样性。是合理地利用这些财富，还是让它从我们的手指间流失掉，这都取决于我们的双手。中国的绿色资源珍贵无比，价比黄金，它的价值体现在经济利益，生态服务，人类健康，生活质量以及精神福祉等各个方面。

　　中国幅员辽阔，自然条件多样，经历了复杂的地理历史变迁，拥有广泛的栖息地类型和多样的物种：从世界之巅的喜马拉雅山到海平面154米以下的吐鲁番盆地；从沙漠、大草原和草甸，到广袤的森林、湿地、海岸和珊瑚礁。

　　中国大约有高等植物35,000余种，分属于353个科和3,184个属（其中有190个是中国特有属）。中国拥有世界上资源最丰富的温带区，是世界上植物多样性最丰富的国家之一。中国也是世界公认的十大"多样性巨丰"国家之一，拥有丰富的脊椎动物及其他动物。

　　中国拥有世界上大约11.5%的哺乳类、13.5%的鸟类、6.7%的爬行类和6.8%的两栖类。中国动物区系的一个特点，是具有很高的地方特有性，也就是说，其中很多种类是中国特有的，在世界其他地区没有分布。

Forests

Natural forests cover 10% of the surface of China. Many types and subtypes are recognized. Forests in the north and on high mountains belong to the mixed conifer type and dominated spruce, hemlock or in some places firs. Forests of the temperate lowlands are dominated by mixed deciduous broadleaf trees. Forests in the south are dominated by sub-tropical evergreen and subtropical monsoon forest types whilst only a few small patches in Hainan, SW Yunnan, SE Xizang and southern Taiwan are clothed in evergreen rainforests. Broad transition zones grade between these major types and different species assemblages make up the respective types in different biogeographical regions of the country. Highly specialized mangrove forests form a narrow fringe in muddy bays along the south coast and around Hainan Island.

Few of the forests in China are in a primary or pristine condition. All have been affected by human activities of logging, thinning, wood collection, burning or replanting. Extensive areas are now reforested with monocultures to produce timber, combat desertification, protect watershed and banks of waterways.

Species richness in the temperate broad leaf and subtropical evergreen forests are the richest of their type in the world and rival the richness of the tropical evergreen forests. Tropical forests are the richest in terms of total flora and total fauna of all terrestrial habitats in China.

Many of China's trees now play a big role in silviculture such as Dawn Redwood (*Metasequoia glyptostoboides*), Chinese fir (*Cunninghamia lanceolata*) and Chine pine (*Pinus massoniana*). Some forest trees have been developed as valuable economic crops such as kiwi fruit (*Actinidia*), chestnuts (*Castanea spp.*), walnuts (*Juglans regia*), gingko (*Gingko biloba*) and the host trees of the valuable lac resin derived from the insect (*Kerria lacca*). Such native timbers as Fujian cypress (*Fokenia hodginsii*), Rosewood (*Dalbergia* spp.) and Agarwood (*Aquilaria* spp.) are among the most precious woods in the world.

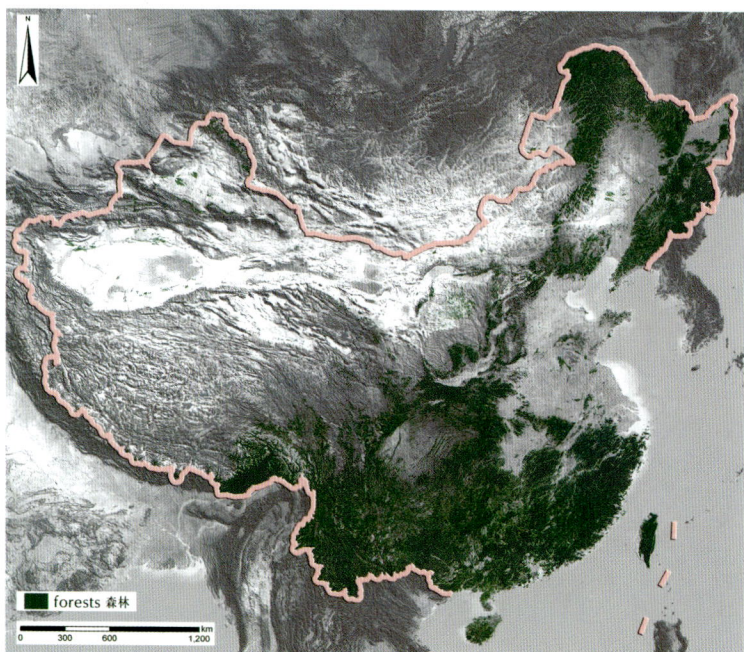

forests 森林

km
0   300   600   1,200

Hainan still harbours some of the densest tropical rainforests in China.

海南仍是中国一些热带雨林的密集港湾。

In temperate woodlands, the spring is a season of many flowers before the deciduous trees develop new leaves that shade the forest floor from sunlight through the hotter summer months.

温带林地，春暖花开后，落叶林生出新的叶片，在炎炎夏季为林地遮阳盖荫。

Fungi serve a vital role in the decomposition ecology of forests.

真菌在林木分解过程中起着重要作用。

Mixed oak woods of northeast China, once rich in deer, pigs and home of tiger and leopards - all now getting scarce and over-hunted.

中国东北混合橡树林，曾富有鹿和野猪，也是老虎和豹子的家，这些动物现在因被过度捕猎变得越来越稀少。

Mists rise over a river through karst limestone.

迷雾升起在一条穿过喀斯特石灰岩的河面上。

森 林

Mixed deciduous broadleaf forest in autumn.
秋季的混合落叶阔叶林。

中国的森林覆盖率为10%，包括多种类型和亚型。北方的高山森林属于针叶混交林，以云杉、铁杉或冷杉为主。温带低地的森林以落叶阔叶混交林为主。南方的森林以亚热带常绿或季雨林为主，只是在海南、滇西南、藏东南和台湾的南部有数量不多的常绿雨林。这些主要森林类型间的过渡区很广阔，不同的物种集中在一起构成了不同生物地理区各自的森林类型。在南部沿海和海南岛，高度特化的红树林则带状分布在泥质海岸。

在中国，真正的原始林很少，现存的森林几乎都经受过伐木、间伐、树木采集、焚烧或种植等人类活动的影响。现在大量的森林都是为了生产木材、治理荒漠化、保护流域和水道而种植的纯林。

和世界相似的森林类型相比，中国的温带阔叶林和亚热带常绿林的物种丰富度是最高的，可与热带常绿林的物种丰富度相媲美。在中国，就动植物总数而言，热带森林是所有陆地栖息地中最丰富的。

中国有很多树种是重要的营林树种，如水杉、杉树和松树；有些森林物种被培育成为极具价值的经济作物，如猕猴桃、板栗、核桃、银杏和生产五倍子的盐肤木；有些乡土树种属于世界上著名的用材种，如福建柏、红木类和沉香木等。

Grassland

The grasslands of Central Asia extend in a huge swathe across northern China and Mongolia whilst alpine grasslands clothe much of the Tibetan Qinghai plateau.

In ancient times, vast herds of gazelles, antelopes and wild horses roamed these pastures, moving with the seasons in great annual migrations. Wolf packs preyed on the weak or sick and cranes and bustards stood tall over the birds and reptiles that also made this their home.

At higher altitudes, yaks and wild sheep grazed the rugged montane meadows, stalked by elusive snow leopards.

These wonderful ecosystems proved rich hunting for the early human settlers. Hunting gave way to pastoral herding. Tibetans domesticated yaks and sheep whilst the Mongolians domesticated horses and cattle. Goats were domesticated further west but gradually were added to the mixed herds. Only the moister, more fertile lands were farmed and for centuries the traditional herdsmen evolved their own nomadic and migratory patterns of using the grasslands in parallel to the herds of wild animals.

It is only in the past 40 years that these old patterns have been abandoned in the name of development. A pattern of sedentarisation, fencing of the grasslands, intensification of stock, expansion of farming, drainage of marshes and extension of road network have all played their part. The revenues generated and the volume of meat and farm produce have indeed been pushed up but at grave cost to the environment and the sustainability of the ecosystems and valuable services they provide.

Less extensive grassland types include reedbeds found in many wetland and coastal sites whilst in the dry valley bottoms of the Nujiang and Lancang rivers in SW Yunnan we can find examples of sub-tropical savannah grasslands. Secondary coarse grasslands (*Saccharum spontaneum*, *Imperata cylindrica*, etc.) now occur where forests have been cleared in tropical and sub-tropical China.

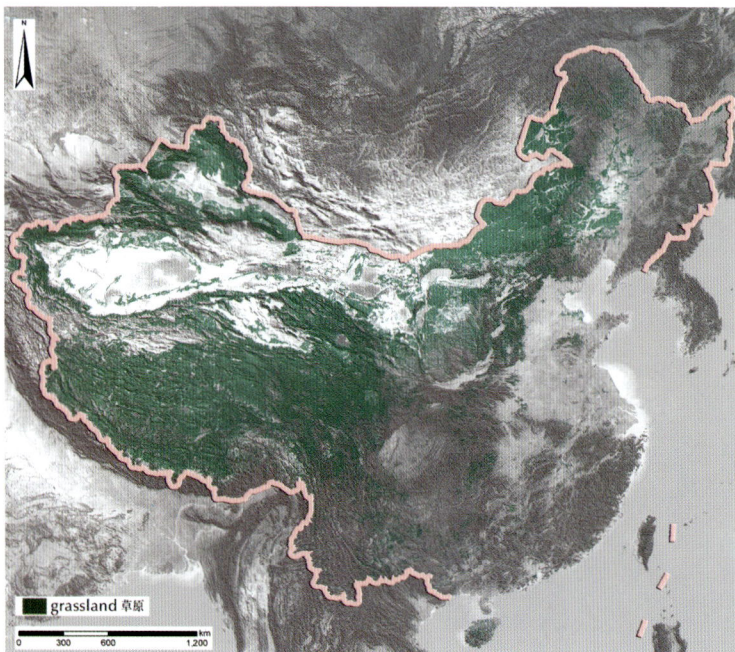

grassland 草原

In summer the lush meadows of Balangshan are a goldmine for botanists.

夏季巴郎山的草场郁郁葱葱，正是植物学家的金矿。

草　原

大草原在中亚大地上如同一条延伸的巨带，贯穿了中国北部和蒙古,高山草地覆盖了青藏高原的绝大部分。

古时候，成群的羚羊和野马在这些草原上漫游，随着季节大量迁徙。狼群在草地上捕食着弱小或伤病的动物; 鹤类和大鸨高高伫立，睥睨着草丛中的其他鸟类和爬行动物。

在高海拔的地区，牦牛和野羊在起伏的草原上徜徉，时刻警惕着出没无常的雪豹。

这些奇妙的生态系统为早期的居民提供了丰富的猎物，后来又成为了畜牧的场所。西藏人驯化了牦牛和绵羊，蒙古人驯化了马和牛。在更往西的地区，山羊也被驯化了，成为家畜中的一员。只是那些降雨丰沛而且比较肥沃的土地才用来耕作。几百年来，牧民们养成了逐水草而居，与野生动物共享草原的游牧习俗。

只是在过去40年里，这些古老的方式才在发展的名义下被废弃，取而代之的是定居的生活方式，和随之而来的围圈草地，集约化养殖，扩张耕作，排干沼泽和交通建设，新的生活方式确实增加了收入，提高了肉类和粮食的产量，但也牺牲了环境、生态系统及其服务功能的可持续性。

小些的草地类型包括湿地和沿海地区的芦苇荡，在滇西南的怒江和澜沧江流域的干热河谷，也能发现亚热带稀树草原。热带和亚热带地区的森林遭到破坏后，会出现浓密的草丛(甜根子草、白茅等)。

Northeast floodplains provide extensive swathes of lush grasslands.
东北平原拥有大片郁郁葱葱的草地。

The high altitude grasslands of Changtang stretch for miles in all directions - one of the wildest and most remote spots on Earth.

羌塘高海拔草原一望无际。羌塘是地球上最荒野、最偏远的地方之一。

Alpine grasslands cover the higher slopes of Changbaishan. There is no boundary fence between China and North Korea.

高寒草地覆盖长白山的较高山坡，中国和北朝鲜会合于此但无国界。

Deserts

Deserts occur in those parts of China with little rainfall. They are characterized by a lack of topsoil and very sparse vegetation. Around the fringes of deserts we can find transition semi-desert formations with increasingly complex and more bushy vegetation before more clearly defined, scrub, grasslands or woodland habitats are encountered where water supply is more regular. About 17% of China can be classed as desert and this figure grows annually as a result of continuing desertification.

Three main types and many sub-types of deserts can be recognized—sandy deserts, stony deserts (Gobi) and the cold deserts of the Tibetan Plateau.

Deserts have rather low productivity as a result of lack of water but many seeds lie in the desert sands and when at last rainfall does arrive the deserts spring into life with a colourful display of flowers. Insects appear as if from nowhere to pollinate the show and desert birds quickly make nests and rear one or two broods of fledglings whilst such bounty lasts and before the normal arid landscape is returned and survival more than reproduction becomes the name of the game.

Wild asses brave the hot and salty conditions with small herds of dainty gazelles. Horny skinned agamid lizards keep a watchful eye out for any unwary flies, Ground jays live in rodent burrows, hardy chukor partridges huddle in small coveys and tough beetles find a mysterious living among the sand and gravel.

Many unique creatures live in the deserts and some of these have remarkable adaptations that enable them to conserve water and survive in such dry landscapes. Most famous of these is the two-humped camel which can store moisture and fat in its dorsal humps that can enable it to live for long periods without food or water.

But it is at night, when the heat of the sun has gone, temperature drop fast and the night creatures emerge from their burrows to scour the bare lands for food scraps. Snakes slither, owls pace about, long legged jerboas bound across the sand on their huge hind legs like miniature kangaroos and desert foxes prick up their big ears to sense the nests of birds or mammals that make their favourite snack.

Many of the desert shrubs have similar water storage devices and have large tubers beneath the ground to keep them alive through prolonged droughts. Many of the shrubs are aromatic. They must protect their precious leaves from grazing animals so lace them with strange strong tasting chemicals as a deterrent. One common species *Artemesia* is sometimes enjoyed as a flavored vegetable by humans. More recently it has been found that the strange flavour can also

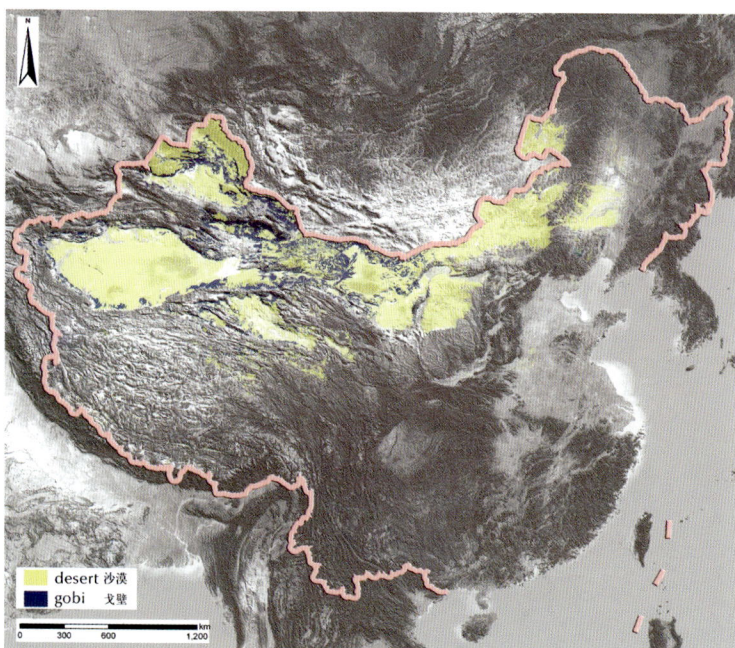

desert 沙漠
gobi 戈壁

0   300   600   1,200 km

serve as a cure for Malaria.

Although the desert yields little harvestable production, the ecosystem is very important for China in that the sparse vegetation is all that serves to bind the loose sand together and reduce the further spread of desert which now threatens to encroach the neighbouring grasslands and even threatens the capital Beijing with increasing frequency of sandstorms. Studying the ecology and propagation of desert plants will be a key to finding solutions to China's desertification problems.

A shower of rain triggers quick flowering and seeding of a desert bush, but dead tree stumps tell of a more favourable former climate.

一阵雨就能使沙漠上的灌木丛迅速开花播种，但死树桩暗示以前的气候可能更宜人。

Sand deserts are spreading in several parts of northern China and sand storms blow many hundreds of kilometers.

沙漠在中国北方的几个地区蔓延，沙尘暴吹刮数百公里。

The camel has been used for centuries as a domesticated beast of burden that can cross the harsh northern deserts.

数百年来，骆驼被驯化成为能载重负荷、能跨越北方寒冷荒漠的家畜。

Agama stoliczkana is beautifully patterned to look like lichens on a rock.

鬣蜥全身有精美的图案，看起来像磐石上的地衣。

Chukor partridges are able to live in some of the harshest habitats in China.

石鸡能在中国一些最恶劣的栖息地存活。

沙 漠

沙漠出现在降雨稀少的地区，这些地区的特点是缺少表土，植被稀疏。在沙漠的周边地区，有过渡性的半沙漠区，其植被更复杂，具有更多的灌木。在水分更充足的地区，能经常见到灌木丛、草地或疏林栖息地。中国的沙漠大约占国土面积的17%，由于持续的荒漠化，这个数字还在逐年增加。

在中国，主要有三种类型的沙漠: 沙质沙漠、石质沙漠（戈壁）和青藏高原的冻原。

由于缺水，沙漠的生产力极低，但沙漠的沙子中有许多种子。雨一旦下来，这些种子就萌发，并开出五彩缤纷的花朵，一些昆虫不知从哪里冒出来，为这些鲜花授粉。这时，沙漠鸟类也赶在干旱时节来临之前迅速筑巢，繁育一二窝后代。这样繁华的插曲转瞬即逝，随之而来的又是漫长无际的干旱，生存又成为荒漠生物的当务之急。

野驴和小群优美的羚羊勇敢地生活在这炎热、恶劣的环境中。长着角质皮肤的鬣蜥瞪大一双警觉的眼睛，时刻准备着捕食不经意的飞虫。地鸦在鼠类洞穴里栖身。勇敢的石鸡紧密地团结在一起。坚强的甲虫神秘地生活在沙砾之间。

沙漠中生存着许多独特的动物，它们具有惊人的适应能力，能以特殊的方式保持水分，使生命在干旱的环境中延续。其中最有名的是骆驼，它能将水分和脂肪贮存在驼峰内，在没有食物和水分的情况下，也能生存很长时间。

到了晚上，当太阳的余热散尽，沙漠的温度迅速下降，夜间活动的动物从洞穴中出来，在赤裸的沙地上寻找食物。蛇在沙地上蠕动，猫头鹰扭转脖子环视四周，长腿跳鼠依靠粗长的后腿在沙地上跳跃，活像小袋鼠。沙狐竖起大耳朵，寻找可以饱餐一顿的鸟巢或哺乳动物。

许多沙漠灌木具有类似的贮水能力，如粗大的地下块茎使它们能够度过漫长的旱季。很多灌木具有芳香气味，它们能在体内生产和积累辛辣的化学物质，使食草动物望而却步，以次来保护自己珍贵的叶子。一种常见的蒿类植物过去时常被人们当作调料食用，但最近发现，这种味道独特的植物具有治疗疟疾的功能。

尽管沙漠生产的可采收生物量微乎其微，但是这个生态系统非常重要，因为稀疏的植被起着固沙的作用，能遏止荒漠化的进一步扩大，而荒漠化的扩张正蚕食着附近的草地，频繁出现的沙尘暴，也正威胁着首都北京。研究沙漠生态和沙漠植物的繁殖技术是解决中国沙漠化问题的关键。

The camel is able to eat dry vegetation in the semi desert.

骆驼能在半荒漠上吃干燥的植物为生。

Mountains

Much of China is mountainous but many of the main mountain ranges are disconnected from one another so that each appears like a biological island, rising above the sea of lowlands between and each mountain range is characterised by the presence of a list of endemic species unique to itself.

The highest mountains are the Himalayas that form the southern fringe of the Tibetan-Qinghai plateau and thus the south west border of China itself. The Himalayas have many peaks over 7000m with Qomolangma the highest in the world at 8,848 m. The mountains contain many glaciers and show distinctly different biology on their north and southern faces. The northern slopes are dry and rugged with little forest and scrubby grasslands only green for a short time each year. The southern faces are by contrast a verdant and rich land, moist from the Asian monsoons. Forests rise to over 5000m and even the alpine zone is lush and bright with a huge variety of colourful flowers. The Chinese border only roughly follows the crests of the peaks so there are a few of the south facing valleys included within its territory and these few valleys add a lot of species that would otherwise not be met within China.

Apart from being the legendry home of the yeti, the Himalayas are the home of red deer, tahr, markhor, different species of marmots and pikas, snow leopards, blue sheep and a host of eagles, pheasants, rosefinches and other birds. The Eastern Himalayas is recognized as one of the biodiversity hot spots of the world.

At the northern edge of the plateau runs the great Kunlun Mountain chain whilst north-west China finds the Tianshan and Altai ranges cutting eastwards with their conifer forests and alpine meadows. In the northeast corner of China rise the Greater and Lesser Xinggan Mountains and on the Korean border the biologically much richer Changbaishan range where Manchurian tiger and Amur leopards still roam.

Leading from the east of the plateau are a series of semi-connected mountain chains - the Hengduan range, Gaoligong range, Minshan range leading to the Qinling Mountains of Shaanxi, Daba range and Shennongjia in Hubei. All these ranges are rich in total species and also homes of many endemic birds.

In southeast China we find another loosely connected series of mountain ranges straddling the borders between Anhui, Jiangxi, Fujian, Hunan and Guangdong. It is in these mountains that the last South China tiger prowls, where the black muntjac browses and the Cabot's Tragopan gives his strange bubbling calls.

In southern China, limestone or karst mountains form a crenillated honeycomb and strange landscape of the Guizhou, Guangxi plateau. Here mysterious landscapes of pillars and cones, giant tiankang dolines and galleries of long caves

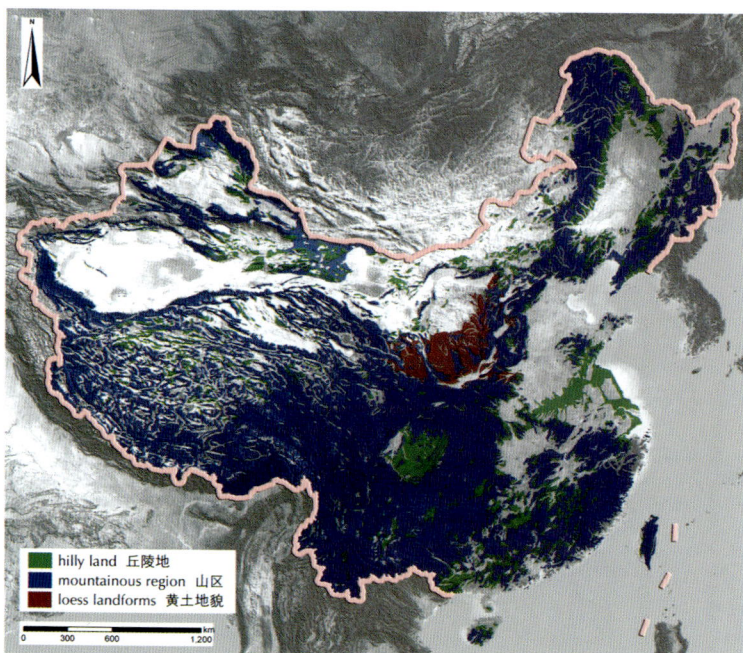

hilly land 丘陵地
mountainous region 山区
loess landforms 黄土地貌

0   300   600   1,200

support a unique biota and rich flora. Protected by the steep terrain, gibbons and rare leaf monkeys maintain a precarious existence. Hornbills flap noisily across the narrow farmed valleys and serpent eagles circle lazily overhead emitting eerie cries.

Towards the tree-line, the forest clings to sheltered slopes as gradually grass and shrubs dominate the vegetation. Still higher, even these no longer survive and bare rock and ice continue to the highest peaks.

靠近森林界线，林木只能在不袒露的斜坡上生长，主要植被逐渐变成草地和灌木。再往高处，草地和灌木都不能生存，只有裸露的岩石，然后就是冰雪至最高峰。

The highest peak in the world. Qomolangma 8,848m from the Chinese side. The other side of the mountain lies within Nepal and is called Everest.

世界最高峰。中国境内的珠穆朗玛峰高八千八百四十八米。珠峰的另一边位于尼泊尔。

The first twittering of waking birds can be heard in the forest as dawn breaks first on the white peak of Meili Snow Mountain.

黎明在梅里雪山的白色峰顶上破晓时，就可听到林中鸟儿们在觉醒时唧唧喳喳地叫唱了。

Purple Asters dazzle. Intense colours and clear skies are typical of the high Himalayas.

紫菀艳丽醒目。浓艳的色彩，晴朗的天空，典型的喜马拉雅山高处景色。

White-browed rose-finch matches the shrub flowers in the high Himalayas.

白眉朱雀在高高的喜马拉雅山上与灌木花卉争艳。

The ridge between Wanglang and Jiuzhaigou is comprised of rugged mountains where eagles glide and blue sheep feed on alpine meadows.

王朗和九寨沟之间的山脊由崎岖蜿蜒的山脉组成，鹰在其上空滑翔，岩羊在高寒草甸上吃草。

山 脉

中国是个多山的国家，但是主要山脉大多彼此隔离，如同分布在平原上的一个个生物岛屿，每个岛屿上都有着自己特有的物种。

喜马拉雅山是世界上最高的山脉，构成了青藏高原的南缘，位于中国的西南边境。喜马拉雅山脉有许多海拔7000米以上的山峰，其中珠穆朗玛峰高达8,848米，为世界之巅。山上有许多冰川，南北两侧的生物特征具有显著差异。北坡干燥、崎岖，森林很少，灌木草地的生长期很短。与北坡形成鲜明对比，南坡植被丰茂，土地肥沃，受亚洲季风影响，空气湿润。森林延伸到海拔高达5000米以上的区域，甚至高山区也青翠欲滴，长满五彩斑斓的鲜花。国境线并不是严格地遵循山脊线，因此有几个山南的山谷是位于中国境内，但就是这为数不多的几个山谷为中国增添了许多物种。

喜马拉雅山脉除了是传说中的雪人的家园外，也是马鹿、塔尔羊、捻角山羊、多种旱獭和鼠兔、雪豹、岩羊、猛禽、雉类、朱雀和其他鸟类的家园。东喜马拉雅山脉还是世界生物多样性的热点地区。

在青藏高原的北端，横亘着昆仑山脉；在西北部，布满针叶林和高山草甸的天山和阿尔泰山延绵向东。大、小兴安岭在中国的东北角崛起，和朝鲜接壤的长白山上生物种类更丰富，东北虎和远东豹仍然在那里出没。

青藏高原的东部是一系列半连接的山环，包括横断山脉、高黎贡山、岷山（延伸进入陕西成为秦岭）、大巴山和湖北的神农架。这些山脉都有丰富的物种，也是众多特有鸟类的家园。

在中国东南部，安徽、江西、福建、湖南和广东的交界处也有一系列连接不太紧密的山脉，这是华南虎的最后家园，也是黑麂和黄腹角雉的乐园。

在华南，石灰岩或喀斯特山脉如同多眼的蜂房，构成了贵州和广西高原的奇特地貌。这个由石柱、石锥、巨大的天坑、落水洞和深邃的岩洞构成的神秘地形养育着一个独特的生物群，有丰富多样的植物。在这陡峭的山地中，长臂猿和稀少的叶猴苟延残喘。在这里，犀鸟振翅高鸣，飞越已被开垦成农田的狭窄的山谷；蛇雕在空中缓缓盘旋，时而发出凄厉的叫声。

Wuzhishan is the highest peak on Hainan but its upper slopes are extremely steep and few people have ventured to the top.

五指山是海南最高的山峰，其上山坡非常陡峭，很少有人挺险到顶峰。

A small salt lake, remains unfrozen in the Tibetan winter.

西藏的一个小盐湖，冬季尚未冻结。

Moist pockets of the eastern Himalayas support good grazing and even small ponds.

喜马拉雅山东部一些潮湿的小区域可以放牧，还有些小池塘。

Hainan contains some of the lushest tropical forests in China.

海南有中国最葱翠的热带林。

Snowcock calling across the mountain valleys.

雪鸡隔着山谷鸣叫。

Golden Eagle flies off after a successful catch of a snowcock.

金鹰成功地捕获一只雪鸡后又飞走了。

Fresh Water and Wetlands

Chinas wetlands cover 6.3% of the country and comprise many types—swamps, peatlands, wet meadows, lakes, rivers, floodplains, deltas, mudflats, mangroves, reservoirs, ponds, and tidal zones. Many of the interior lakes of the Tibetan Plateau are saline to various degrees and even though there is little replenishment from rainfall, There is a constant trickle of water from the melting of glaciers.

Chinaís wetlands provide great economic, ecological and social benefits, and so their conservation is very important. These services include flood mitigation, water storage, climate regulation, water purification, erosion control, land formation and creation of scenic and relaxing environment. They also provide homes for a wide array of plants and animal species including birds, fish, amphibians, crustaceans and many plants and animals that are used by humans for food, including some very valuable shrimp habitats.

It is a little-known fact that China's peatlands store more carbon that all the forests of China combined. Preserving such peatlands from drying or burning is a great global contribution to ameliorating climate change.

Yet despite the obvious importance of wetlands in China, these critical habitats are being degraded and lost at a frightening pace as a result of pollution, drainage and lowering of water tables.

Indeed several large lakes have already dried up such as Lop Nur, Manas and Juyanhai.

It is urgent that greater efforts are quickly taken to reverse this trend and restore vital 'lungs' to the Chinese landscape.

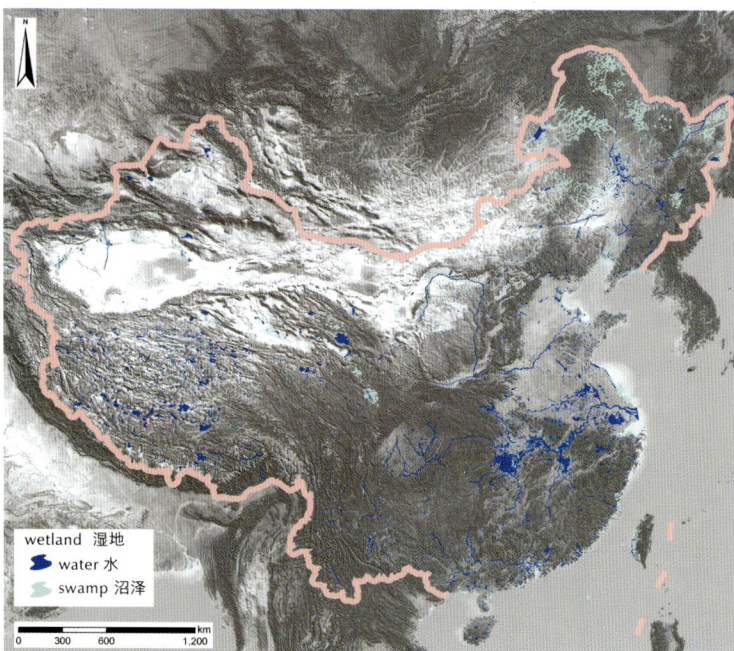

wetland 湿地
water 水
swamp 沼泽

0   300   600   1,200   km

A herd of Tibetan gazelle feed on the lush green pastures beside a plateau lake.

一群藏原羚在高原湖边郁郁葱葱的牧场上食草。

As wetlands are drained and converted into farmlands, the sun sets for the last time on may wetland species.

湿地正被排水转化为农田, 五月的夕阳西下, 最后一次照在湿地物种上。

An endless sea of grass and water marks the great lake of Dongting in Hunan Province.

湖南的洞庭湖草水相映, 似浩淼无边的海洋。

Otters have become very scarce in China as a result of river pollution and over-fishing. In former times otters were trained like cormorants today to catch fish for their masters.

由于河水污染和过度捕捞，水獭在中国已非常稀少。过去的水獭犹如今天的鸬鹚，被训练成为主人捕鱼的助手。

Horsetail′s sprout like miniature trees from the water surface.

问荆的新芽像微型树木一样，从水面上升出。

Flock of tundra swans flying over the reedbeds of Yeyahu nature reserve close to Beijing.

北京附近的野鸭湖自然保护区，一群小天鹅飞越芦苇丛。

淡水和湿地

湿地面积占中国国土面积的6.3%，湿地类型有多种，包括沼泽地、泥炭地、湿草甸、湖泊、河流、漫滩、三角洲、滩涂、红树林、水库、池塘和潮汐带。青藏高原上的许多内地湖泊是咸水湖，但盐度不尽相同。青藏高原降水稀少，这些湖泊更多地是依赖冰川融化的涓流来补充水量。

湿地为中国提供了巨大的经济、生态和社会利益，因此湿地保护十分重要。湿地的服务包括：控制洪水、水源涵养、调节气候、水质净化、控制水土流失、淤积土地、提供观光和旅游资源。湿地也为众多的动植物提供了生存空间，其中包括鸟类、鱼类、两栖动物和甲壳类，以及很多为人类食用的动植物。

一个鲜为人知的事实是，中国的泥炭地所贮存的碳量大于所有森林贮存的碳量之和。对泥炭地进行保护，防止其变干或燃烧，能为改善全球气候变化做出重大贡献。

尽管湿地具有显著的重要性，但是由于污染、开垦和水位下降，这些脆弱的栖息地正在迅速减少，而且日渐退化。

事实上，有几大湖泊已经干涸，例如罗布泊、玛纳斯湖和居延海。加大力度逆转这种趋势，恢复中华大地之"肾"，已经迫在眉睫。

Inviting pool wells up as a spring after traveling through underground caves.

清凉引人的池水流经地下洞穴后涌出成温泉。

Purple irises form a magnificent show in a marshy hollow.

紫鸢尾花在空旷的沼泽地显得十分美丽。

A pair of kingfishers rest near their nest in a riverbank burrow.

一对翠鸟，在河岸边的巢穴小憩。

Marine Habitats

China's seas range from the cold temperate seas of the north, the semi-enclosed Bohai Sea, the oceanic east coast and all the way down to the tropical coral reefs of the Xisha and Nansha archipelagos.

In the northern seas one can find marine mammals such as seals together with a wide range of sea gulls and other birds whilst below the water surface live great fisheries of tuna and mackerel, squid and bass.

In the south are colourful coral gardens teaming with colorful fishes, strange mollusks, dolphins, sharks and marine turtles.

Deep in the ocean giant whales sing their mournful songs and make their annual passage between polar waters.

China derives a great bounty from its seas in the form of fish, squid, mollusks, crabs, shrimps and sea slugs. Many of the most treasured Chinese dishes depend on this sustained supply—abalone, napoleon wrasse, sea horses and sharks fin.

海洋栖息地

Nemo the clown-fish lives among the sea anemones.

尼莫小丑鱼在银莲花中生存。

中国海洋栖息地的跨度极大，从北方的冷温带海洋、渤海湾、东海岸一直延伸到西沙和南沙群岛的热带珊瑚礁。

在北方海域，可以发现诸如海豹之类的海洋哺乳动物、海鸥及其他鸟类，水面下有金枪鱼、鲭鱼、墨鱼和鲈鱼等巨大的渔业资源。

在南方海域，珊瑚礁五彩缤纷，华丽的鱼类、奇特的软体动物、海豚、鲨鱼和海龟穿梭其中。

在远洋海域，巨鲸唱着哀怨的歌曲，每年往返游弋于南北极之间。

鱼、软体动物、蟹、虾和海参都是大自然给中国的优厚礼物。许多著名的中国菜肴就是用鲍鱼、濑鱼、海马和鱼翅作原料。

Flock of black-headed gulls following boat.

一群红嘴鸥跟随着鱼船。

Marine turtles live in the coral reefs and warm seas of southern China. They nest on beaches of oceanic islands and a few places on the south coast.

海龟居于珊瑚礁和暖热的中国南海。它们在岛屿海滩上及南方沿海少数地方筑巢。

Coral reefs provide homes for the widest range of living species in the marine world.
珊瑚礁为海洋世界里最广泛的生物物种提供家园。

Starfish along the sandy shoreline.

海星点缀沙滩。

The Blue-ringed Angel Fish is one of the most colourful in the reefs of the South China Sea.

蓝环天使鱼是中国南海珊瑚礁中色彩最艳丽的鱼。

Diversity of Peoples

Because human culture and natural species are closely related, the diversity of people in China and their different attitudes towards and uses of wild species creates its own dimension of biodiversity.

The Chinese population is predominantly one ethnic race—the Han. But even within the Han there are marked regional and cultural differences, crops, food preferences, clothing styles and attitudes towards biodiversity. Farmers grow different crop varieties, raise different animals and harvest different wild species in relation to regional preferences and local conditions. Sichuan farmers like their food spicy, Guangdong folk cannot bear chilies. Or as the ancient saying goes - east spicy, west sour, south sweet and north salty.

In addition China recognizes 55 other ethnic minorities, many with their own language, representing several races and practicing a diversity of religions. Many ethnic groups maintain strong spiritual or cultural attitudes towards sacred mountains, lakes, forests or special species.

Buddhists of Tibet and other parts of China respect all life and shun the killing of animals whenever possible. The Muslims of the north-west abhore the eating of pork and can only eat other flesh if killed in haram tradition.

The Dai minority in the south of Yunnan protect sacred 'longshan' or holy hills above their villages in respect for the ancestors and in the belief that such forests safeguard water supply and protect them from disease.

The Mongols along China's northern frontiers are great horsemen. In summer they live in tents and tend their herds of domestic animals. They are also good hunters and have hunted for meat and fur for centuries.

But for all their differences, the different ethnic groups in China all share certain underlying Buddhist and Taoist attitudes of belief in harmony between man and nature and a respect for living creatures. These are important beliefs that can well form the basis for developing new conservation ethics in China.

Dai villages are comprised of neatly arranged wooden stilted houses with tile roofs. Animals sleep below the main house.

傣族村庄由一排排整齐的瓦顶木屋组成。动物睡在屋子底层。

Tibetan farmers in their summer camp near Lake Qinghai.

藏族农民在青海湖夏令帐篷附近。

Mongolian sacred site, decorated and piled high.

蒙古族圣地，装饰浓烈，堆得很高。

民族多样性

由于人类文化和自然物种密切相关，因此中华民族的多样性、人们对自然的不同态度及利用，也必然创造出生物多样性的不同层面。

中国人口以汉族为主。但是，由于地区和文化的差异，同样是汉族，他们耕作的作物、对食物的偏好、衣着风格和行为方式，还有对生物多样性的态度都表现出显著不同。农民种植不同的作物品种，饲养不同的家畜，根据不同地区人们的偏好及地方条件，采用不同的野生动植物。四川人喜欢麻辣食品，广东人却受不了辣椒。用句古老的俗语来说就是：东辣西酸，南甜北咸。

除汉族外，中国还有55个少数民族，其中许多少数民族都有自己的语言，代表不同的种族，信奉不同的宗教。对许多少数民族而言，山峦、湖泊、森林和某些特定的物种具有特殊的精神和文化内涵。

西藏及其他地区的佛教徒尊重所有的生命，尽量不杀生。在西北地区，根据传统，伊斯兰教徒不吃猪肉，只吃按伊斯兰传统宰杀的其他肉类。

云南南部的傣族对村庄边上的"龙山"或"神山"进行保护，以表达对先祖的尊重，他们相信山上的森林能蓄涵水源，并保佑他们安康。

中国北方边境地区的蒙古人主要以牧马为生。夏天，他们居住帐篷，饲养家畜。他们捕猎技术良好，几个世纪以来一直靠打猎以获取肉类和毛皮。

中国各民族各群体尽管有各种差异，但都基本信仰佛教和道教，尊重人与自然间的和谐，尊重众生。这些信仰很重要，可成为中国新的保护理念的基础。

Prayer wheels in Buddhist monastery in Qinghai.

青海佛寺内的转经轮。

Buddhist Thanka style painting still thrives in western China.

佛教唐卡绘画艺术仍然在中国西部蓬勃发展。

Tibetan farmers in Jiuzhaigou take a break at harvest time.

九寨沟的藏族农民在收获时小憩。

Mongolian boys learn their horsemanship looking after the family herds.

蒙古男孩在学习马术以照顾家畜。

Minority girls from Guangxi Province.

广西的少数民族少女。

Cicadas and scorpions - all for sale in China food stalls.

蝉和蝎子－在中国小食品摊位上都有销售。

Divination with Animal and Plants

A happy Dai minority girl in colourful garb, clambers among the trees to collect the edible flowers of the tree Maoodendron igniferum.

快乐的傣族少女身着色彩缤纷的服饰，攀爬于树木之间，收集 *maoodendron igniferum* 可食用的花卉。

Divination with plant and animal products dates back to more than 5000 years ago. Its importance reached zenith during the Shang Dynasty (1600 - 1100, B. C.) when decisions had to be made with it for almost every major activities, such as military campaign, agriculture, astronomy, hunting trips and interpretation of dreams.

The two oldest methods of divination, the Chinese prophecy in the earliest forms, were the 'Bu' which used turtle shell and shoulder blade of animal (cattle, sheep etc), and 'Shi' which used the herb *Achillea sibirica*.

Bu used only the plastron or breastbone of the turtle. A hole was bored in the center. Then a hot spike was inserted in the hole and cracks formed along the edge. Bu scholars would read the form of cracks and inscribe their interpretation on the plastron. This was in fact the earliest Chinese characters.

Shi used 50 stems of the herb *Achillea sibirica* , according to a certain permutation and combination, drew a group of digits to speculate the future of persons or events by checking with the prescribed types of fortunes in Ba Gua symbols of Taoism.

卜筮与动植物

用动植物产品占卜可追溯到5000多年前。在商代（公元前1600年-1100年）尤为盛行，原来殷商人举凡祭祀、征伐、农事、气象、狩猎、旅游，甚至释梦等事，都必须先占卜。

"占龟"和"祝蓍"是两种最古老的占卜方法，也是我国远古时代预言本的最早形式，占龟使用龟甲或牛羊等动物的肩胛骨，而祝蓍则使用蓍草。

求卜兆使用龟甲或动物肩胛骨，先钻后灼，根据龟甲或兽骨裂出的兆象，预言吉卜。并将占卜结果和事后灵验与否刻在甲骨上。这事实上也就成了中国最早的文字。

祝蓍是将蓍草按照一定程式排列组合，求得些数字对照八卦以推测人事吉凶。

Achillea sibirica was the chosen plant for fortune divination since ancient times.

蓍草自古以来就被用于算命占卜。

The breast bones of terrapins were inscribed in ancient times with fortune interpretations. These are the earliest examples of writing in China.

龟甲是古代占卜的工具。甲骨文是中国最古老的文字。

PART TWO
The Importance of China's Biodiversity

Benefits Derived from China's Biodiversity

China's great biological diversity is immensely precious. Variety has intrinsic value. Compare two libraries each containing 10,000 books. One library only contains 10 titles with shelves of identical books, the second containing 10,000 different titles is much more valuable, especially if some are rare and out of print titles. It contains more knowledge, more opportunity to guide innovation and development. More biological richness—what we refer to as 'biodiversity'—means more options for developing new crops, crop improvements, discovering new medicines and maintaining more robust, resilient and adaptive ecosystems.

Biodiversity is vital for human survival. It forms the basis for the critical life-support systems upon which we all depend. Biodiversity underpins the continued provision of ecological goods and services, such as clean air, regular supplies of clean water, conservation and renewal of fertile soils, pollination of crops and natural vegetation, as well as the maintenance of forests, wetlands, oceans and a wide variety of other ecosystems essential for a well-functioning planet. Where we see biodiversity lost and degraded, we see increased desertification, higher incidences of natural disasters, disease epidemics and reduced crop yields. Biodiversity is so ubiquitous we take it for granted. Because its market values are not always immediate and direct, we underestimate its importance for economic and social stability and development. Biodiversity especially impacts the poor, because they are often directly dependent on it for their livelihoods. Various efforts to put a value on China's biodiversity and its services all conclude that this exceeds gross national productivity of all productive sectors combined. Increasing the understanding and awareness of the role and importance of biodiversity is key to making sure we conserve it for present and future generations.

When we lose species from an ecosystem, the system becomes less and less stable and may collapse just as a house will collapse if you continue to cut out the supports for firewood. Healthy natural ecosystems are the essential firm foundation on which to build secure economic development. In China's case, good water supply is the key to continued economic growth. Deforestation, overgrazing, draining of wetlands, siltingand pollution of water bodies, destruction of species that serve as pollinators and seed distributors all affect water supply reducing overall flow but increasing seasonal flood peaks. Receding glacier and its impact on livelihood of local people. Maintaining healthy forests and grasslands is thus a vital component of human welfare and continued development and conserving the component species of those ecosystems is vital for their health and function.

Black-naped oriole utters melodious calls all through the early summer in woodlands and orchards of most of eastern China.

初夏，在华东大部分地区的林地和果园都可听到黑枕黄鹂悠扬的叫声。

第二章
中国生物多样性的重要作用

中国生物多样性带来的
价值利益

生物多样性的价值巨大，而且差异本身就具有价值。若将生物资源比作图书馆，现有两个图书馆，分别藏书 10,000 册，第一个图书馆只有 10 种图书，书架上摆放着相同的书；第二个图书馆有 10,000 种不同的书，有很多是珍贵或绝版的书。显然，第二个图书馆所包含的信息量更大，更有可能为创新和发展提供帮助。生物丰富度（也就是我们所说的"生物多样性"）越高，意味着有更多的原材料来培育新作物、改良作物品种和发现新的药物，来更好地维持生态系统的活力、结构和应变能力。

生物多样性对人类的生存至关重要，它构成了我们赖以生存的重要的生命支持系统的基础。生物多样性确保了生态系统产品和生态服务的持续供应，包括清洁的空气、不间断的清洁水源、土壤肥力保持和更新、作物和天然植物的授粉，森林、湿地、海洋等众多生态系统的维持，这些生态系统是地球发挥其正常功能所必需的。生物多样性丧失和退化的后果，是荒漠化不断扩大，自然灾害更加频繁，瘟疫肆虐，粮食产量下降。由于生物多样性几乎是无所不在的，人们往往熟视无睹。因为它的市场价值并不总是能立即和直接地显示出来，因此人们往往低估它在经济社会的稳定与发展上的重要性。生物多样性对贫困人口的影响尤其严重，因为他们的生计通常是直接依赖于生物多样性。大量研究结果表明，中国生物多样性及其服务价值超过了各生产行业总生产能力之和。因此，加强对生物多样性的作用和重要性的了解和认识是我们采取保护行动的关键。

当一个生态系统内的物种开始消失时，这个系统的稳定性将会逐渐降低，甚至最终完全崩溃。如果人们把一座房屋的梁一根一根拆下来当柴烧，总有一天房屋会坍塌，其中的道理是一样的。因此，健康的自然生态系统是经济安全发展的基石。就中国而言，充裕的水源是经济增长的关键。毁林、过度放牧、开垦湿地、水体淤塞和污染、授粉和散布种子的物种的消失都将减少河流的总径流量，加剧季节性洪水的危害，从而影响到水源的供给。因此，维持健康的森林和草地是人类福利和持续发展的重要组成部分，而保护生态系统的物种又是维持生态系统功能和健康的关键。

Treefrogs range surprisingly far north. Enduring the severe cold of Heilongjiang winter, these amphibians emerge to enjoy the summer warmth.

树蛙令人惊讶地能分布在远北地区。承受过黑龙江省的冬季严寒后，这些两栖动物出来享受夏天的温暖。

Camelia is the genus of tea but many species have been brought into gardens because of their showy flowers.

茶花属于茶树类，但许多物种因为花美而被种植在花园。

Takin - with a long mournful face and horns like an African Wildebeest.

羚牛. 脸部长长的，看似悲哀，犀角像非洲野羚。

The moist forest supports a wealth of orchids. Some are used for medicine, some collected for horticulture but mostly left where they are to beautify the natural vegetation.

潮湿森林中有许多兰花。有些被用于医药，有些被园艺收集，但大多数留在原地美化自然植被。

The valuation of ecological services of Dongting Lake

Dongting Lake, located in Hunan and Hubei is one of China's largest freshwater lakes. As a result of recent agricultural impounding and natural sedimentation, the Dongting Lake which was formerly the second largest freshwater lake after Poyang Lake and measured about 6,000 square kilometers in 1949 had dropped by 1983 to only 2625 square km. In recent years, strengthened protection of the lake region and restoration of farmland to lake has increased the area of the lake back to a total of 3968 square kilometers.

A team of the Chinese Academy of Sciences Subtropical Agro-ecological Institute, headed by research fellow Wang Kelin, after 9-year study found that the Dongting Lake wetlands provided ecosystem services with a total value of 209,700 million yuan, of which the value of the water regulation accounts for 78% and material production accounted for only 2.3%. The recent expansion of the lake provides an increase in the value of ecological services of 7,090 million yuan.

In their assessment of the Dongting Lake ecosystem services value, the group suggested that Lake's leading ecological functions should be recognized as "a very important Yangtze River flood control". The main auxiliary ecological functions as a "middle and lower reaches of the Yangtze River water Balance of the key functions of the District" and "Wetlands of International Importance of rare and migratory birds wintering habitat" and "an important agricultural and sideline production base in the fishing industry."

Interestingly, in 2004, the Dachang Chuang Nanjing University published another study also showing that the value of Dongting Lake wetland's ecological services was mainly in the flood storage, water supply and climate regulation but there was a considerable differences in values. According to this study the direct and indirect use values amount to only 8,072 million yuan; of which flood storage is valued at 3,712 million yuan, accounting for 45.99 percent of the total.

Cormorants are tethered awaiting their turn on the fishing boats. Each bird is tied to a line and each time it catches a fish it is hauled back to the boat, gives up the fish in return for a small reward.

鸬鹚拴系于渔船上，等待着轮到它们去捕鱼。每只鸟颈上用线绑上，每捕到一条鱼，就被拉回渔船，把鱼吐出，以换得一点小小的回报。

洞庭湖的生态服务价值

位于湖南、湖北之间的洞庭湖曾是中国第一大淡水湖。由于现代的围湖造田，以及自然的泥沙淤积，洞庭湖面积由最大时的约6000平方公里骤减到1983年的2625平方公里，从而退居鄱阳湖之后，成为第二大淡水湖。近年来加强了对湖泊区域的保护，实行退耕还湖，使湖区面积有所恢复。现在天然湖泊面积2625平方公里，蓄洪堤垸和单退堤垸高水还湖扩大湖泊面积1343平方公里，总共3968平方公里。

以中科院亚热带农业生态研究所王克林研究员为首的课题组经过9年研究发现，洞庭湖区湿地生态系统服务功能价值总和为2097亿元，其中价值最大的水调节功能占78%，物质生产功能仅占2.3%。特别是退田还湖后，洞庭湖湿地生态系统发生了一系列变化，生态服务功能价值增加了70.9亿元。在评估了洞庭湖区湿地生态系统的服务功能价值后，课题组提出，洞庭湖的主导生态功能应为"长江流域极重要的调蓄滞洪区"，主要辅助生态功能为"长江中下游水域生态平衡的重要功能区"、"国际重要湿地和珍稀候鸟越冬栖息地"和"重要农副渔业生产基地"。

有趣的是，南京大学学者庄大昌2004年发表的另外一项研究表明，洞庭湖湿地生态服务功能价值主要表现在调蓄洪水、供水和调节气候等方面的价值，但经济价值的数额却有相当的差异。据他的研究，洞庭湖湿地的生态服务功能价值、直接利用价值和间接利用价值总和为80.72亿元，其中调蓄洪水的价值为37.12亿元，占45.99%。

Villagers collect lotus leaves on Dongting Lake. The leaves are used for wrapping meat in local cooking style.

村民在洞庭湖收集荷叶。荷叶在当地被用于烹饪时包裹肉类。

Even the smallest fish find a market.

即使是最小的鱼也有人买。

Sustainable Harvest

Biodiversity is very productive. Since the area of natural habitats is many times larger than the area of China used for agriculture we can guarantee that total productivity is far higher. But whilst agriculture produces predominantly plants grown as food for humans, nature produces a great variety of products of which only a few are harvested as food. Even so the harvest of wild fish, edible wild plants, fungi, game meat is still substantial. The Dai minority people in Xishuangbanna for instance recognize and harvest over 300 wild plants as food.

More than 1500 species are harvested as traditional medicines. Chinese Traditional Medicine is used by hundreds of millions of Chinese with a domestic value and export market of several $ billions per year.

Much of what man cannot eat can be eaten by his domestic animals. So biodiversity provides the food that nurtures the huge herds of cattle, yaks, horses, sheep, goats and supplements food for pigs, ducks, chickens, rabbits, bees and other livestock and fisheries required to feed 1.3 billion human inhabitants of the country.

We can list many other products harvested from nature such as resins, essential oils, honey, brooms, poles, timber, thatch and firewood.

Forest wood is still the main fuel for more than half the population of China providing a harvest of about 300 million cubic metres of firewood annually. Little is paid for this commodity as it is harvested almost freely from nature, but the alternate cost of using other fuels if wood was not available would be crippling to a huge number of Chinese peasants.

In the future, biofuels will play a growing role in providing green energy for transport and other uses. These are also the direct products of biodiversity.

Even the large agriculture and aquaculture sectors are entirely dependent upon biodiversity for their raw germ plasm and all new developments and improvements rely on biodiversity as the source of genetic materials. One could argue that these human production sectors owe a 'royalty' to biodiversity just as a car producer has to pay a royalty to the designer of each part of the car.

Even the oil, gas and coal that we use to fuel most of our transport, factories and heating is the product of fossil biodiversity. We are cashing in on long-term deposits derived entirely from the production of living organisms. These are not renewable sources and we should use them frugally as a one-off bonus.

Honey-making is a good village industry, compatible with biodiversity conservation and dependent on a multitude of wild flowers.

养蜂采蜜是一项良好的村镇工业，与生物多样性保护原则相符，而且它依靠多种多样的野生花卉。

Soft-shell turtle makes futile bid for freedom but likely meets the cooking pot very soon.

软壳龟努力争取自由，但可能难逃被锅中蒸煮的命运。

可持续利用

生物多样性具有很高的生产力。由于天然栖息地的面积要比中国的农业用地大好几倍,因此其总生产量肯定更高。但农业是以生产粮食为主,而在大自然的众多产品中,被人类食用的只占很少的比例。即使如此,人类依然采集了大量的野生鱼类、食用植物、真菌和猎物。例如,西双版纳的傣族采集用作食物的野生植物多达300多种。

有15,000多个动植物种被用于传统医药。千百万中国人使用传统中药,每年在国内外市场上的销售价值达到几十亿美元。

有许多人类不能食用的物种可以用来喂养家畜或家禽。生物多样性提供了饲养大量牛、牦牛、马、羊、山羊、兔、蜜蜂、鱼、猪、鸭、鸡和其他家畜所需要的食物,满足着13亿居民对肉类的需求。

人类还从大自然采集许多其他产品,如树脂、精油、蜂蜜、扫帚、柱子、木材、茅草和薪柴。

森林仍然是一半以上中国人的主要能源,每年供应大约3亿立方米的薪柴。薪柴几乎是大自然的赠品,若没有薪柴而使用其他燃料,将使大量中国农民不堪重负。

将来,生物质燃料将在提供绿色能源上发挥越来越重要的作用。自然,生物质燃料也是生物多样性的直接产品。

规模庞大的农业和水产业的种源、新品种的开发与改良也依赖生物多样性提供遗传物质。可以认为,人类的生产行业应该向生物多样性支付"专利费",就如汽车生产者需要向汽车零件设计者支付专利费一样。

就连我们交通、生产和取暖所不可或缺的石油、天然气和煤炭也是生物多样性化石的产品,它们是生物界千万年来累积的产物,而我们却坐享其成,在短期内将其挥霍殆尽。对于此类不可再生资源,我们应该将它们当作一次性红利来珍惜。

Edible lilies grow beside the roads.

长在道旁的食用百合。

Villagers around Changbaishan supplement their income by collecting edible mushrooms in the forest.

长白山周围的村民们在森林中收集食用菌以补充收入。

Field eels ready for market. A favourite dish in Sichuan.

准备好要上市的鳗鱼。这是四川人最喜好的一道菜。

Service Values　生态服务价值

The ecological services provided to China by its biodiversity form the basis for agriculture, aquaculture, secure constructions and human health. Major components include:

生物多样性所提供的生态服务是农业和水产业发展，安全建筑和人类健康的基础。主要的生态服务如下：

Climate Regulation

Vegetation cover helps ameliorate climate in several ways. Plants create a cool land surfaces that attracts and holds rain clouds, fog and dew. Clearing vegetation creates hot land surfaces which leads to heat thermals and speeds the dissipation of clouds. Vegetation replenishes the oxygen supply on which we depend and removes carbon dioxide from the atmosphere. Too much carbon dioxide results in loss of the protective ozone layer and to global warming by limiting the loss of heat from the planet surface to outer space. This is the so-called "greenhouse" effect.

气候调节

植被有助于改良气候。植物创造一个吸引和保持雨云、雾和露水的凉爽的土地表面。植被遭到破坏后，地表温度更高，热量会加强雨云的消散。植被能从大气中吸收二氧化碳，补充我们赖以生存的氧气。太多二氧化碳会破坏具有保护功能的臭氧层，也会阻止地球表面的热量被散发到太空中去，形成所谓的温室效应。

Climate warming is changing the frequency and severity of tropical storms.

气候变暖加剧了热带风暴的发生频率和强度。

Nutrient cycling

Plants need small quantities of minerals such as phosphorus, potassium, calcium, and magnesium to grow properly. In mineral rich floodplains or slopes of volcanoes these nutrient minerals may be in abundance but on old lands with heavy rainfall they can become leached out and soils become acid and infertile. Yet in nature such weak soils often exhibit luxurious vegetation. This is because the forest plants are extremely efficient at nutrient recycling. As soon as living tissue falls to the ground a host of decomposers break down the tissues and the released nutrients are immediately absorbed by fungal mycorrhizae living on the roots of other plants. Thus fertility is preserved. When natural vegetation is disturbed nutrients are lost and washed away.

In addition, nitrogen is taken from the atmosphere and taken into living tissue by plant roots and the CO_2 levels in the atmosphere are reduced and oxygen levels replenished by the photosynthetic activities of green plants. All these functions help to preserve and healthy air and fertile soils for human use.

营养物质循环

植物的正常生长需要微量的矿物质，如磷、钾、钙和镁。在洪漫滩或火山山坡上，矿物养分很丰富，但是在降雨量高的长期耕作的地区，矿物养分可能会流失，最终土壤地变成贫瘠的酸性土。然而，在自然界，这种贫瘠的土地上通常长着丰茂的植被。这是因为森林植物中的营养物质循环效率非常高。当生物组织落到地上，许多分解者将它分解，所释放的养分迅速被生长在其他植物根系上的真菌所吸收，肥力借此得以保留。若天然植被受到破坏，养分将流失掉。此外，植物根系能将从大气中吸收的氮，固定到生物组织中；绿色植物还能通过光合作用吸收大气中的二氧化碳并释放出氧气。所有这些功能都有助于保持健康的空气和肥沃的土壤，供人类使用。

Creation and conservation of soils

The global value of natural soil formation and protection has been estimated at 17 trillion U.S. dollars per annum (Costanza 1997). Biodiversity is fundamental in the formation and maintenance of soil structure, as well as its retention of moisture and nutrient levels. Without this function farm production would collapse.

Many kinds of organisms, from trees to microorganisms are involved in the process of soil formation. Soil formation involves the break-down of large rocks into finer particles. Lichens and the roots of trees and other plants achieve this slow and gradual process and improve water penetration. By taking up raw nutrients from the decaying rocks and converting them into living tissue, the plants contribute organic matter to the soil mixture, through leaf litter and other decaying tissues. Microorganisms, fungi, and invertebrates break down and recycle these organic materials to create healthy soil conditions. Some plants, such as legumes, develop specialized roots where symbiotic microbes live and absorb atmospheric nitrogen, "fixing" the nitrogen into the soil making it more fertile for other plants. Plant root systems can also absorb and bind with toxic minerals, such as iron and aluminum and remove them from the soil.

The soil building community is easily disrupted by excessive human activity. The loss of biodiversity—through vegetation clearing, monocropping, use of too many chemicals, land engineering, wetland drainage, contributes to the leaching of nutrients, accelerated erosion of topsoil, salinization of floodplains, and laterization. All result in serious loss of fertility.

Natural vegetation comprises many species co-adapted to live densely together—deep rooted with shallow rooted—and work as an efficient team. Man's simple agricultural systems are far less efficient at soil protection. For example, a hectare of tropical rainforest rarely loses more than 1 ton of soil annually. However, when the forest is leveled and planted with various crops, the erosion increases drastically. If the forest is replaced with dense vegetation like a coffee plantation, the hectare loses between 20 and 160 tons, whereas if it is replaced with field crops, the patch can lose more than 1,000 tons annually.

形成和保护土壤

全球土壤形成和保护的价值估计为每年17万亿美元。估计中国占全球价值的15%。生物多样性是形成和保持土壤的基础，因为它能保持水分和养分，否则，农业生产将崩溃。从树木到微生物的多种生物体，都参与了土壤形成。土壤形成源自大岩石被分解成小颗粒。地衣和树木及其他植物的根系完成了这个缓慢、渐进的过程，并提高了土地的透水性。植物从分解的岩石中吸收初始养分，将它转化成生物组织，并通过落叶和其他衰退组织增加土壤的有机质。微生物、真菌和无脊椎动物将有机物质分解，创造出健康的土壤条件。豆科植物能生长出供共生菌生存的根系，吸收大气中的氮，将氮固定到土壤中，增加土壤肥力，从而有利于其它植物生长。植物根系也能吸收和固定诸如铁和铝之类有毒的矿物质，将它们从土壤中除去。

过度的人类活动很容易破坏形成土壤的生物群落。植被破坏、单一作物耕作、过量使用化肥和农药、土地整理和开垦湿地都会造成生物多样性丧失，进而引起养分流失、土壤的表土侵蚀、漫滩盐碱化和红土化。这些后果都将严重导致土壤肥力丧失。

天然植被包括许多相互适应的物种，根系生长深度不同的植物紧密地生活在一起，如同一个紧密团结的团队。简单农业系统的土壤保护效率低下。例如，一公顷热带雨林每年流失的土壤不超过一吨。然而，若森林被清除并种上经济作物，土壤侵蚀将大大增加。若天然林被诸如咖啡之类的人工林所取代，每公顷土地的土壤流失量为20～160吨；而若被庄稼所取代，那么每公顷土地的土壤流失量将达到1,000吨／年。

Dung beetles and other creatures speed up the recycling of animal waste into the soil.

屎壳郎能加速动物粪便的降解，促进营养物质循环。

Tree roots cling tightly to the limestone blocks. The combination of roots, algae, erosion and bacteria help form new soil among the leaf litter.

树根紧抱石灰石块。树根、藻类、土蚀及细菌结合一起，使零乱落叶形成新的土壤。

Selaginella ferns on the floor of Laojunshan.

老君山森林地面上的蕨类植物－卷柏。

Gilled underside of sulfur mushroom in Laojunshan - much favoured for food.

老君山上金黄色的蘑菇是一种美味。

Hydrological Functions

In forest areas, tree roots dig into the soil allowing deeper water penetration during times of rain and increasing the soil's water holding capacity. The level of carbon in the soil derived from rotting vegetation also increases its water holding capacity. Vegetation also uses up water from rain and in the soil and releases this back directly into the atmosphere by transpiration.

In total these effect reduce run-off during rainy periods thus limiting soil erosion and floods but by holding water back in the soil 'sponge' allows a more continuous supply of filtered clean water for human uses. The efficiency of hydropower generation is far higher, the more evenly the supply can be given throughout the year.

The good turf of natural grasslands and the peaty soils of China's many wetlands perform similar functions in the non-forested regions of China.

The increasing damage due to both floods and droughts in recent years in China is a result of a combination of the natural vegetation 'sponge' having been destroyed through deforestation and overgrazing together with the blocking or silting of lakes and rivers that serve as the drains to release flood waters. This silting is also caused by loss of natural vegetation.

Mixed natural forests, perform these hydrological roles better than artificial monoculture plantations.

Fresh mountain streams churn down the steep mountainsides.

生机勃勃的山涧沿陡峭山坡翻腾而下。

水文功能

在林区，树根能深入到土壤中，在下雨时，雨水能渗透到更深的土壤里，从而提高土壤的保水功能。植被死亡腐败后，留在土壤中的碳也能提高土壤的保水能力。植被吸收雨水和土壤的水分，并通过蒸腾作用将它们释放到大气中。

这些影响将减少雨季的水土流失，从而限制土壤侵蚀和水灾，将水分保留在土壤这个"大海绵"中能为人类持续供应清洁的用水。另外，如果一年中不同季节的河流径流量越均匀，水力发电的效益就越高。

在中国缺乏森林的地区，天然草地的完整植被和很多湿地的泥炭质的土壤发挥着类似上述森林的功能。近年来，由于毁林、过度放牧破坏了天然植被的"海绵吸水"功能，以及湖泊与河流堵塞或淤积，水灾和旱灾所带来的危害日益加重。河道淤积通常也是因天然植被遭到破坏的结果。

天然混交林的水文功能比人工纯林的水文功能强。

Clean water from the healthy ecosystems of northwest Xinjiang.

源自新疆西北健康生态系统里的纯净水。

Coastal Protection

Coral reefs around China's southern coastal fringe and around the South China Sea islets break the surge energy of wave action and tsunami's, helping to protect coastlines and coastal properties from erosion and tidal damage.

Mangroves stabilize silt both protecting the coral from silt smothering but also serving as storm shelter to villages and towns from typhoons. Coastal grasses and herbs bind loose sand and mud to form new lands. Along China's east coast the coastline grows seawards by several metres per year. These are extremely valuable new lands for agriculture and real estate. They are in danger of being lost back to the sea if either the vegetation is destroyed or sea level rises too fast as a result of global warming.

海岸保护

中国南部沿海和南海小岛周围的珊瑚礁消耗了波浪和海啸的能量，有助于保护海岸线和沿海财产，使之免受土壤侵蚀和潮汐的破坏。

红树林固定了淤泥，一方面保护了珊瑚免受淤泥的覆盖，另一方面也保护了村庄和城镇免受台风的袭击。沿海草本植物起着固定松散的沙子和泥土，形成陆地的作用。在中国东部沿海，陆地每年以几米的速度向海里延伸。这些是农业和不动产的新珍贵用地。若植被遭到破坏或因全球变暖海平面上升过快，这些土地又将面临重新失去的危险。

Reed beds stabilize coastal estuaries and provide excellent wildlife habitat.

芦苇能保护河口的堤岸，并为野生动物提供良好的栖息地。

Saline vegetation along the east coast, helps capture mud and create valuable new lands.

东海滩涂的盐生植物能帮助固定泥沙，加速土地淤积。

Maintaining Ecological Balance

Primates perform important roles in the ecosystem both as plant predators eating leaves, shoots and soft fruits but also as important fruit dispersing agents, swallowing seeds that pass through the digestive tract to be planted within a nice package of natural fertilizer at a good distance from the parent tree. So useful is this behaviour that evolution has designed some fruits to be specifically dispersed in this way rather than by any other animals. *Litchi*, *Garcinia*, *Citrus* and *Gnetum* have rinds that are laced with spiny hairs or distasteful oils. These deter squirrels from eating into them but can easily be peeled off by more dexterous primates. The advantage of a primate over a squirrel is obvious. Squirrels sharp teeth can easily gnaw into the rich seeds, whilst primates seek only the sweet flesh and discard the hard seeds unharmed.

Carnivores also perform an important regulatory role. Small carnivores like frogs, wasps, spediers and dragonflies do a largely unseen buthugely important labour of killing caterpillars and mosquitos and other creatures that would otherwise threaten human health and destroy our crops. In parts of southern China, farmers noticed that they get better Citrus crops from trees that have nests of weaver ants. You can even buy nests of ants in the market to put in your trees. The ants attack insect pests that might threaten the health of the tree and even clean the leaves so they are more efficient at trapping solar energy.

Hawks, snakes and foxes control the small rodents. In parts of China where people have collected too many snakes as food, people notice that they now get plagues of rodents. Lacking the natural control agents, farmers have to use poison to kill the rodents but his in turn is dangerous to people, fish and other species within the ecosystem.

At the top end of the carnivore scale—wolves, leopards and tiger control the numbers of deer, antelopes and wild pigs. Once the carnivores are wiped out the numbers of large herbivores get out of balance with the vegetation. This leads to overgrazing or diseases which lead to under-grazing, forest fires, thicket formation and other problems.

Wolves can still be found on the high plateau despite persecution for their valuable fur.

尽管盗猎猎獗，但是在青藏高原上依然可以寻觅到狼的踪影。

维持生态平衡

灵长类在生态系统中发挥着重要的作用，一方面，它们取食植物的叶子、嫩芽和软果实，另一方面，它们取食的果实中的种子通过消化道，和粪便（天然优质肥料）一起排出，往往落在离母树很远的地方，发挥着传播种子的重要作用。这种传播方式非常有效，很多植物经过长期的演化，已经形成了专门依靠灵长类来传播种子的结构。荔枝、藤黄、柑橘和买麻藤的果实具有长满刺状物或难闻气味的外壳，能阻止松鼠的咬食，但是却能被更灵巧的灵长类轻易地剥开。与松鼠相比，灵长类具有明显的优势。松鼠锋利的牙齿很容易咬破种子，而灵长类只吃果肉，将完整的种子扔掉。

食肉动物也发挥着重要的调节作用。小型食肉动物喜欢吃青蛙、黄蜂、蜘蛛和蜻蜓，起着杀死毛虫、蚊子和其他危及人类健康和毁坏庄稼的动物的作用。

在华南的部分地方，农民注意到，如果柑橘上有黄蚁，那么柑橘的收成就会更好。你甚至可以从市场上买个蚂蚁窝放在树上。蚂蚁能攻击危害树木的害虫，甚至还能清理树叶，使树叶能更有效地进行光合作用。

鹰、蛇和狐狸控制着小型啮齿动物的数量。在中国的部分地区，由于捕食了太多的蛇，人们现在正在为啮齿动物所困扰。由于缺少天敌，农民必须用药物来杀死啮齿动物，但是反过来，这又给人类、鱼和生态系统中的其他物种带来了危险。

食肉动物狼、豹和老虎控制着鹿、羚羊和野猪的数量。一旦食肉动物灭绝，大型食草动物与植被之间的平衡将被打破。结果，食草动物或者过度啃食植被，或者因疾病而大量死亡，进而又将导致啃食不足、森林火灾、林下植被层过密以及其他问题。

类似的问题也发生在淡水和海洋环境。只要人类扰乱了自然的平衡，资源采集超过了资源补充的限度，都会出现类似的问题。

Spotted Sika deer browse on tree leaves in the dappled woodlands.

梅花鹿在斑斓的林叶间啃食嫩叶。

The Siberian piebald toad Bufo raddei occurs in NW and NE China and is noxious to predators.

花背蟾蜍在华西北和华东北有分布，对猎食它的动物有毒。

Wild pigs are widespread in the forests and scrub of China. They are important in sorting through the seeds, fruits, carrion and worms of the forest floor. Their digging activities cultivate the soil and affect which plants will form the new forest.

野猪在中国的森林和灌木丛中很普遍。从其（通过拱动）挑选种子，水果，腐肉和森林地上的蠕虫的意义上而言，它们对森林系统意义重大。它们的挖掘举动使土壤松弛，还影响着哪些树木将形成新的森林。

Natural biological controls

Spiders, ants, frogs, dragonflies and insectivorous birds provide an important service in controlling insect pests that would otherwise destroy our crops and threaten human health. The alternate approach of spraying insecticides is expensive, dangerous to health and causes insects to evolve tolerance to the chemical applied.

At the micro level, natural fungi and bacteria perform similar controls that limit many plant diseases and other pest micro-organisms.

天然生物控制

蜘蛛、蚂蚁、青蛙、蜻蜓和食虫鸟类在控制虫害上发挥着重要的作用，如果这些动物消失，我们的庄稼将被毁，人类健康也要受到威胁。虽然喷洒农药能控制虫害，但成本很高，对健康有害，而且昆虫也会逐渐产生抗药性，使农药不再有效。

在微观层面，天然真菌和细菌能发挥类似的生物控制作用，能控制许多植物疾病和其他有害微生物。

Natural pollination

Bees and other insects provide a free service in pollinating a vast range of domestic fruit trees, silvicultural trees, horticultural flowers and vegetables. Loss of these natural pollinators would be a disaster to some economic sectors. In tropical regions bats and some birds also perform these pollinator roles.

天然授粉

蜜蜂和其他一些昆虫为许多果树、营林树木、园艺花卉和蔬菜提供免费的授粉服务。失去这些天然授粉者对一些经济部门来说将是一个灾难。在热带地区，蝙蝠和一些鸟类也扮演着传授花粉的作用。

Whip scorpion is a primitive arachnid and useful small carnivore of the forest floor.

鞭蝎，是生活在森林地上的一种原始蛛形纲动物，也是一种有着其重要功能的小型食肉兽。

Bees perform an invaluable service of pollinating flowers and fruit trees.

蜜蜂免费为花卉和果树授粉。

A large spider prepares his silky trap for butterflies and other insects.

大蜘蛛以丝设陷阱，来捕捉蝴蝶及其他昆虫。

Hummingbird hawk moth hovers to pollinate Lantana flowers.

蜂鸟蛾在 Lantana 花上授粉。

Praying mantis feeds on a variety of insect pests.

螳螂是多种害虫的天敌。

Tourism and Recreation

Recreation is an essential part of human welfare, good for health, a chance to recharge the body and mind and therefore leads to higher productivity. Outdoor tourism is a huge part of Chinese recreation and this is one of the fastest growing industries in this dynamic country.

Where formerly people rarely had the chance to travel and then did so timidly in a guided tour stopping only for a few moments to take some photos at a famous landmark, today hundreds of millions of Chinese are making trips further and deeper into the beautiful countryside and willing to spend high prices on clothes, accessories, transport and entrance fees.

Foreign tourists flock to China with even fatter wallets to see the unique and fantastic scenery and wildlife. More than 200 million visitors come to China each year. The total tourism industry is currently valued at 100 billion RMB per annum. A huge industry of transport communications, hotels, restaurants, souvenirs and destination facilities has arisen and continues to grow. A large part of this is dependent on biodiversity.

旅游和娱乐

娱乐是人类福利的基本组成部分,有益于心身健康和维持良好的状态,从而有利于提高工作效率。户外旅游是中国娱乐业的一个重要组成部分,也是这个中国的一个快速增长的行业。

以前人们很少有机会出去旅行,后来亦步亦趋地跟着导游去商场购物,然后在著名的景区拍几张照片。如今,数亿中国人到美丽的景区旅游,也愿意在衣服、饰物、交通和门票上慷慨解囊。

也有更多的国外游客来中国参观独特、迷人的风景和野生动物。每年到中国参观的游客超过2亿人次。目前,旅游业的年总产值为1000亿元人民币。一个包括交通、通讯、酒店、宾馆、纪念品和旅游点设施的庞大产业已经形成,并在持续增长。这些产业在很大程度上也依赖于生物多样性。

Off the beaten route, visitors get a closer encounter with nature. But impacts may be too great if visitor numbers are large.

沿循未被走过的线路,游客可更贴近自然。但如果游客数目很大,对自然的影响也可能会加大。

Tourists enjoy a rafting trip. Safety has to be a high priority but white water rafting is available for the more adventurous.

游客享受泛舟之旅。安全是应予以最优先考虑的事。但对更愿冒险的游客们，白水泛舟也已可能。

Nature reserves need information centres to explain to visitors what can be seen and give meaning to what they will observe.

自然保护区应设立信息中心，向旅客们解释参观的是什么，其含义是什么。

Research and Genetic Resources

Biodiversity has yielded the raw genetic resources for all domesticated species now used for agriculture, fisheries, forestry and animal husbandry. It continues to yield new genes that can be crossed or engineered into such domestic races and varieties.

The variety of biodiversity under direct human domestic propagation and selection is referred to as agro-biodiversity. But the process of maintenance and continued improvement of these stocks requires trying out new variants, bringing in new adaptive qualities and adapting to changing conditions and needs. This is increasingly important as we enter a period of fast global change when conditions of weather may vary quite rapidly.

There are still a huge range of natural species and products that have never been researched. Genes that can give higher yields, genes that can give stocks disease resistance or ability to survive in broader environmental extremes or under greater stress.

Many relatives of domestic species can be crossed with domestic varieties are of particular value, so are those species that are harvested and eaten or used by local people but not yet domesticated on any commercial scale.

Two species of enormous importance in providing security to the Chinese diet are wild rice and wild soya beans. A newly discovered population of wild rice was used to cross new genetic vigour into agricultural varieties and found to increase yields and lower water dependence on both traditional rice varieties and modern high yield varieties.

Soya bean has been cultivated in China for 5000m years. Several wild species related to cultivated soya have huge potential in improving the rather narrow genetic range of cultivated varieties.

Another important are is in the search for new medicinal compounds in nature. For thousands of years people have used plants as medicines. Today we can analyze and synthesize many of the active compounds or develop similar molecules that may have good medicinal effects. A new area of search is among the poisons of stinging and biting animals. Many such compounds have been found to have good use but a huge area of research remains un-tackled and to throw away our biodiversity before exploring it would be to throw away many of our future development options.

Chaenomeles japonica bears smart red flowers that have made this bush a gardeners favourite.

日本海棠的美丽红花，使之成为园丁最喜欢的园中植物。

An effective new modern drug for treatment of malaria was recently developed in China by extraction from the wormwood shrub *Artemesia annua*. Tea made from the leaves of this plant had been listed as a traditional remedy for malaria and other diseases for centuries.

The highly valued morel fungi grow well in Laojun Mountains of NW Yunnan and offer the chance of an alternative livelihood for local people around the nature reserve.

价值昂贵的羊肚菌在滇西北的老君山易生易长，为自然保护区周围的当地人民提供了很好的替代生计。

研究和遗传资源

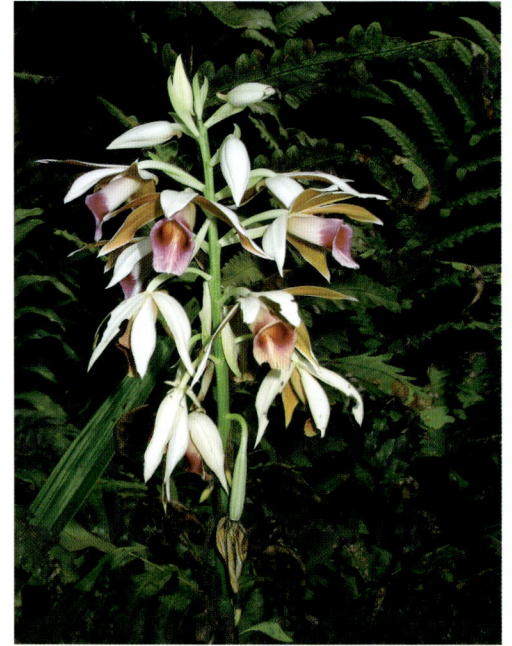

Some of the ground orchids are highly prized and fetch high prices among collectors. Reserve staff have to be vigilant to prevent collectors from removing these plants from their reserves.

一些地生兰花价值很高，人们不惜以高价收藏。保护区员工必须随时警觉，以防保护区的兰花被盗。

生物多样性不但为现代农业、渔业、林业和畜牧业的所有驯化物种提供了原始的遗传资源，而且也提供了新的基因用于培育新的家用品种。

人类直接繁殖和选择的生物的多样性称为"农业生物多样性"。维持和不断改良这些原料的过程需要尝试新的变种，引入新的适应性品质，以便适应不断变化的环境。随着我们进入一个快速全球化变化的时代，气候条件也可能迅速变化，"农业生物多样性"对我们越来越重要。

许多驯化物种的亲缘物种可与驯化品种杂交，因而具有特殊的价值。同样，那些被当地人采集或食用但尚未被进行商业化驯养的物种也具有特殊的价值。

对中国食物保障有巨大作用的两个物种是野生水稻和野生大豆。最近发现的野生水稻被用于与农业品种杂交，结果发现野生水稻的遗传活力能增加农作物产量，降低对水的依赖，对传统水稻品种及现代高产水稻品种均有改进作用。

中国种植大豆已有5000年的历史。有多种野生大豆品种与栽培品种互有关联，有潜力用于增进栽培品种相对狭窄的基因。

另一重要方面是从自然界寻找新的药用化合物。人们已经有几千年使用药用植物的历史。今天，我们能分析和合成许多具有良好药效的活性化合物，或开发出类似的分子。一个新的研究领域是动物蜇、咬时排出的毒物。已经发现，许多这类化合物具有重要用途，但许多研究领域尚未触及。在对它们进行探索之前将我们的生物多样性抛弃，不啻于把我们未来的发展机会抛弃掉。

最近，中国研究开发出一种新的对治疗疟疾有效的药物，该药物是从植物黄花蒿中提取出的。几个世纪以来黄花蒿的叶子就被用来泡茶，来治疗疟疾及其他疾病。

Peonies were favoured by Chinese emperors and many varieties domesticated in imperial gardens.

牡丹向来为中国皇室所喜爱，许多品种在皇家园林有种植。

Pansies smile in unison. Beijing and other cities are spending large sums of money on urban greening efforts.

紫罗兰笑口同开。北京和其它城市不惜花费大量资金进行城市绿化。

Spiritual Values

Can one imagine Chinese culture without its biodiversity? The scenery, plants and animals of China have been the inspiration for so much of the stories, poetry, art and imagery for thousands of years. The natural heritage is every bit as important and valuable as such cultural relics as the Great Wall, temples and palaces.

Chinese landscape painting is a unique cultural concept and differs from the foreign landscape art in that it does not reproduce real natural landscape, but rather represents an ideal landscape for aesthetic and cultural reflection.

Landscape painting emphasizes the harmony and affinity between man and nature.Spring, summer, autumn and winter, the light of morning and evening, wind, rain and snow, mountains and rivers each show a different aspect and all embody the meaning of life. Through the landscape we see the life of inner strength and spirit of movement.

Chinese people get this idea of landscape from their primitive religion. Worship of the mountain spirit, the water goddess has gone through a long historical process. When Gods ruled society, they were hidden unseen in the mountains and rivers. Thus landscape is the mother of the gods. The height of mountains, depth of water; width of the hills and vastness of lakes all implied the infinite. mystery of the universe. To show the gods due worship and awe, people depicted the image of mountains and rivers for respect and pay tribute. The myth of God or painting of god inspired landscape paintings.

The philosophical ways determined the Chinese definition of nature and religion gives life and spirit to the philophical ways. Heaven, earth and I live together, heaven, earth and I are one. Zhuangzi unveiled the mysterious relationship between man and nature, man thus started to relate to heaven and earth spiritually. The benevolent finds happiness in mountains and the wise in waters

For centuries Chinese monks have sought quiet meditation and spiritual peace among the wild hills and mountains. The land is dotted with sacred lakes, forests and mountain peaks including the famous ten senior Sacred Mountains revered by Buddhist and Taoist monks—Omeishan, Huangshan, Taishan, Wutaishan, Putuoshan, Jinhuashan, Huashan, Hengshan, Songshan and Nanyehuangshan. There are many other lesser sacred mountains.

Many of China's 55 minority races have deep religious associations with hills, lakes, rocks and even single trees. In SW China villagers protect 'longshan' or holy hills for many centuries for religious and practical reasons. Koreans climb to worship the scared Changbaishan. Tibetans make painful pilgrimages to sacred lakes and glaciers.

Visitors and local people place sticks and stones as sign of respect to rocks and trees.

旅客和本地人民把棍棒和石头放置在此表示对岩石和树木的尊重。

精神价值

若没有生物多样性，难以想象中国文化会是什么样的？几千年来，中国的风景、植物和动物一直都是故事、诗歌、艺术和绘画的灵感来源。这些自然遗产就如长城、庙宇和宫殿等文化遗产那么重要和珍贵。

中国山水画

山水画是中国人所特有的一个文化概念。中国人的山水画不同于外国人的风景画，它不是再现自然景观，而是通过对自然景观的表现，赋予自然以文化的内涵和审美的观照。

天人合一人，与自然的亲和，是山水画的基础。春夏秋冬，朝暮夜昼，风霜雨雪，山水表现出了不同的面貌，体现了生命的意义。山水和人一样具有内在的生命运动和精神力量。

中国人关于山水的观念，源于原始的宗教。对山灵的崇拜，对水神的敬畏，人们在这种崇拜和敬畏之中经历了漫长的历史过程。在神统治的社会里。一切神灵无不隐匿于山水之中。山水是神灵之母。山之高，水之深；山之广，水之渺，蕴涵了天体宇宙的无限奥妙。为了表示对山水神灵的崇拜和敬畏，人们刻画山水神灵的形象，用于祭祝或瞻仰。神话或神画启发了山水画的独立成形。

哲学方式决定了中国人关于自然的定义，宗教则为哲学方式增添了生命和精神。天地与我并生，天地与我合一，庄子打开了人与自然之间的屏障，人开始与天地精神往来。"仁者乐山，智者乐水"。

"天下名山僧占多"。几个世纪以来，中国的和尚习惯于在山川中坐禅和修身养性。中华大地点缀着许多圣湖、圣林和神山，其中包括被佛教徒和道士所敬重的十大闻名的神山：峨眉山、黄山、泰山、五台山、普陀山、九华山、华山、恒山、嵩山和衡山。另外，还有许多知名度较低的神山。

中国的55个少数民族中，有许多民族的宗教都与山、湖泊、岩石和树木有关。在中国的西南地区，村民们出于宗教和实用的原因，将"龙山"或"神山"保护了好几个世纪。如藏民长途跋涉，不辞辛劳地前往神山和神湖进行朝拜；朝鲜族登上神圣的长白山去敬神。

China is the home of gold fish. Centuries of selection from common carp have resulted in many colourful and decorative varieties.

中国是金鱼之家。几百年来对鲤鱼种类的选择，产生了许多色彩缤纷极具观赏性的品种。

The famous Li river scenic area near Guiyang town attracts millions of tourists a year to see the strange landscape of the karst stacks.

贵阳附近著名的漓江风景名胜区，每年吸引了数百万游客来参观其奇特的景观——溶岩堆叠。

The sandstone pillars of Zhangjiajie in Hunan Province form part of a World Heritage site and attract more than 2 million visitors per year. A cable car can carry visitors to the top of the peaks.

湖南张家界的砂岩柱，是该世界遗产景点的组成部分，每年吸引2万多游客。电缆吊车可把旅客载上峰巅。

Using the Power of Nature

We can see clearly the value of ecological services provided by healthy habitats, but is it necessary to maintain high biodiversity to ensure such services ? The answer is yes. Complexity equates to stability, adaptability and higher efficiency.

Moreover, animals form an integral and essential part of healthy ecosystems. Ecosystems are regulated and balanced by the impact of animals that serve as selective predators of plants, from tiny caterpillars to huge elephants; pollinators of plants from bees and moths to birds and bats, seed dispersers of plants such as monkeys and ungulates and assist in the break down and decomposition of plants and dead animal material. Carnivorous animals are necessary to keep in check some of the faster breeding and aggressive herbivores and different sized carnivores deal with different sizes and types of prey.

Ecosystems are dynamic, constantly changing and balancing things out. As conditions change, so the composition and balance of the species mix responds accordingly. If we have a period of 20 warm winters we would see an increase of the warm-adapted plants in the forest composition, the new forest would retain its efficiency without human intervention. If one species becomes too numerous, its predators or diseases will increase and pull it back down to normal levels. If a species suffers a setback and has high mortality due to some disease or climatic extreme, then its numbers are reduced below natural levels, those predators that seek it out will switch to alternate prey because the effort of finding the now rare form is less rewarding. It will enjoy a period of low predation and recover its lost ground.

Each time there is a storm, a landslide or even when a large tree tumbles with a crash to the forest floor dragging other trees in its wake and creating a gap in the forest canopy, there is a chance for new species to colonize and fill the gap or damaged habitat. These events provide opportunities for change and enrichment of the site and allow closer adaptation of ecosystem to prevailing conditions.

This ability to heal and change means that natural ecosystems can perform their services of climate regulation, water control, nutrient recycling and soil creation and protection without human inputs. There is no need to plant, prune, weed or irrigate. There is no need to spray fertilizer or pesticide. The service is free and sustainable in a way no artificial habitat can duplicate. Natural forests are age staggered, there are populations of immature and juvenile plants ready to take their place in the ranks and replace the ageing and dying senile members. Moreover, natural vegetation is layered into stories—emergent tree crowns that

Natural conifer forest in NE China.

中国东北的天然针叶林。

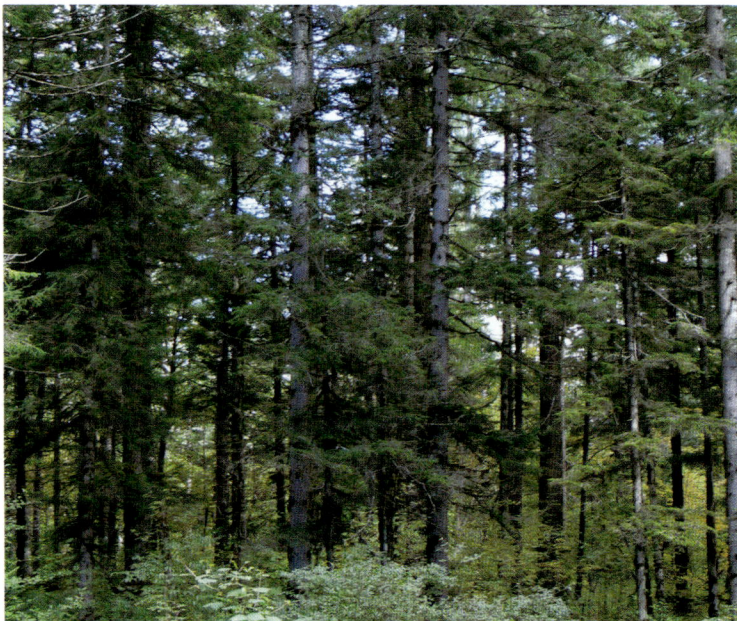

stand tall of the main canopy; middle canopy of smaller trees and saplings beneath the main crowns and a dense under-storey of herbs and shade tolerant shrubs or grasses at ground level. This ensures that almost every ray of precious sunlight is captured by living leaf surface and converted through photosynthesis into living production.

Nature does it better and nature does it for free. Man's simple ecosystems may generate more saleable products but they cost a lot in terms of labour and maintenance and rarely come close to natural ecosystems in delivery of ecological services. For instance field measurements show that rates of soil erosion under rubber plantations are one hundred times greater than under natural forest.

Flocks of Red-billed starlings feast on trees of strangling figs.

绞杀榕上的累累果实是红嘴椋鸟的丰盛大餐。

利用大自然的力量

我们能清楚地看到健康栖息地所提供的生态服务的价值,但是要保证这些服务的供应是否就必须保持丰富的生物多样性? 答案是肯定的。生态系统的复杂性实际上就等同于稳定性、适应性和更高的效率。

此外,动物是健康生态系统的基本的、不可或缺的组成部分。生态系统通过动物的影响进行调节和维持平衡,动物选择性地取食植物,小到毛毛虫,大到大象,都有自己的取食对象。从蜜蜂和蛾类到鸟类和蝙蝠都是植物的授粉者;猴子和有蹄类动物不仅是种子的散布者,还有助于破碎和分解动植物尸体。食肉动物有利于控制繁殖过快的动物和侵略性食草动物。大小不同的食肉动物也会捕食大小不同的猎物。

生态系统是动态的,通过不断变化达到平衡。当条件发生变化时,物种的组成和平衡也会相应地发生变化。如果我们连续有 20 个暖和的冬天,那么我们将会看到,森林里喜温植物的比例会相应增加,这样,改变后的森林仍将保持它的生态服务效率,而无须人类的干涉。如果某种物种数量太多,它的捕食者或疾病将增加,把它拉回正常水平。如果某种物种的种群数量由于疾病或极端气候的影响而低于自然水平,那么,由于捕食该物种变得更加困难,捕食者将转而去寻找别的捕食对象。经过一段时期以后,该物种的数量又将恢复原来的水平。

每出现一次暴风雨,每发生一次泥石流、山崩,每一株大树连带四周的树木轰然倒地,把林冠砸出一个缺口,都为新的物种创造了侵占该空隙或受损栖息地的机会。这类事件为林地条件的改良创造了机会,同时也使生态系统更适应当前的环境。这种恢复和改变的能力意味着自然生态系统能在无需人类干预的情况下提供调节气候、保留水分、循环养分、形成和保护土壤的生态服务。不需要种植、修剪、除草或灌溉,也不需要施用化肥或农药。这些服务是免费和持续供应的,但却不是人工栖息地所能复制的。天然林的年龄参差不齐,有众多未成熟的年幼植物准备取代老化和垂死者。此外,天然植被呈层状分布,露生层立于冠层之顶,中间冠层由更小的树木和幼树构成,地表层由草本植物和耐阴灌木构成。这保证了几乎每一丝宝贵的阳光都被叶片利用,通过光合作用,生产原材料。

自然生态系统能很好地提供生态服务,且是免费的。简单人工的生态系统可生产更多可销售的产品,但是这些产品需要大量劳动投入,且在提供生态服务方面远不及天然生态系统。例如,实地测量表明,橡胶林的土壤侵蚀速度是天然林的100倍以上。

Chinese bulbul feeding on fruits. Birds distribute seeds in their droppings and help to spread scrub and forest.

取食果实的白头鹎。鸟类吃下的种子随着粪便排出,有利于灌木和森林的扩展。

More precious than gold

One commodity of wildlife was regarded as even more precious than gold in ancient times. That was the feathers of the common kingfisher. The plumage is a beautiful cobalt blue with iridescent green as you turn the feathers in the light. These were regarded as the finest ornaments for decoration of the filigree silver head-dresses worn by fine ladies of the imperial courts or in Beijing Opera. So great was the price paid for good feathers that the bird was hunted close to extinction and even neighbouring countries of Southeast Asia could not meet demand. As far away as Indonesia special laws had to put in place to protect kingfishers from this fashionable trade. Today, the kingfisher has recovered and now faces pollution rather than persecution as its greatest threat. You can still find old pieces of kingfisher feather jewelry in the antique markets, but expect to pay a high price.

比黄金还珍贵

古时候，普通翠鸟的羽毛被视为比黄金还珍贵。普通翠鸟的羽毛呈美丽的钻蓝绿色，在光亮中发出金属光泽。贵妇淑女们进王宫或到剧院听京剧时都佩戴装饰精美的金银头饰，而普通翠鸟的羽毛就是头饰上的最上品点缀。有钱人愿付高价购买精美的羽毛，因此该鸟被极度捕杀，几近灭绝。翠鸟羽毛如此热门，甚至从邻近的东南亚国家进口还供不应求。印度尼西亚就已制定出专门法律，以保护翠鸟不要因这种对时髦的追求而进行的贸易而灭绝。今天，翠鸟数量已经恢复，目前它们所面临的最大威胁是污染，而不是被杀害。在古玩市场上您仍可以找到有翠鸟羽毛装饰的珠宝件，但得准备付出高价。

Qing Dynasty head-dress made of silver and kingfisher feathers.

翠钿——清代用银和翠鸟羽毛制成的首饰。

The common kingfisher is one of the most colourful birds in China. During the Ming and Qing Dynasties the blue feathers of kingfishers were used with filigree silver to create bright jewelry for court ladies broaches and headpieces. The bird became very rare but has now largely recovered.

普通翠鸟是中国鸟类中色彩最缤纷的一种。在明、清两朝，翠鸟的蓝色羽毛与花丝银一起被用来打造胸针和头饰等珠宝以供宫廷仕女们享用。该鸟数量一度变得稀少，但现已基本恢复。

PART THREE
Review of China's Special Wildlife

Pandas

China's most famous wildlife is of course the giant panda. Giant pandas are found in an arc to the north and west of the Sichuan basin ranging from the Qingling mountains of Shaanxi provenance to the Baishuijiang region of southern Gansu and the Minshan, Qionglaishan and Liangshan ranges of western Sichuan. In prehistoric and even historic times giant pandas ranged over a much range of China. But climate change and pressure from humans has driven them back to these lost few refuges. One could say they have their back against the wall. The wall being the inhospitable heights and cold regions of the Tibetan plateau. But it is those very heights that prevent the moisture of the monsoons from leaving the valleys of west Sichuan and that is why the region is proverbially damp and cloudy and why the bamboo undergrowth thrives so well in the sub-alpine forests. Bamboo is the staple diet off the pandas and they eat little else. Much has been made of the threat to pandas by the bamboo's habit of mass flowering over large areas; followed by the dying of the adult plants and the many years it takes for the new seedlings to replace the original stock. But that is exactly the pattern within which the panda has evolved to cope.

Such flowering events may indeed cause starvation and death to some pandas but to others it stimulates emigration. They walk cross valleys or cross high mountain crests to find new valleys where the bamboo has not flowered. This dispersal stage may be vital for both maintaining selection by weeding out the weak and unhealthy pandas and also creating the necessary levels of out-breeding required to ensure genetic health of the panda population.

Pandas will cross farm lands not because they are coming down to villages is to beg for food but because they are travelling to the other side of the valley in search of more bamboo. It is a misconception to imagine these animals require rescuing except in extreme circumstances. Most of the so-called "rescued" pandas are perfectly healthy and should be allowed to continue their wandering, or at a maximum, given a good meal and then put back in the wild to find their way again.

China has over indulged in captive breeding with several hundred pandas now in captive institutions. Even though breeding success has improved and 30 to 40 young panda cubs are born each year in captivity in China; we have still not returned any successfully back into the wild. Although the captive population is now capable of sustaining itself, the overall panda breeding program has been a drain rather than a benefit to the wild population.

Efforts at panda conservation must continue to protect and safeguard the wild populations and to design a system of nature reserves that are safe for pandas and yet providing enough corridors for pandas to migrate when they need to from one population to another. The number of panda reserves has therefore been doubled in recent years and almost all of the remaining panda habitat is now contained within protected areas.

Pandas are also heavily protected by law. There is a death penalty for anyone found killing or selling pandas

or their parts. Some traders were executed a few years ago when they were found to have a stash of pandapelts. But even this does not prevent many pandas from dying in the snares that are set for musk deer and other wildlife by hunters. Gradually the wild panda population is on the rise and may number as many as 1600 once more. But the large earthquake of 2008 has damaged the pandas habitat and new surveys are needed to find out the impacts on the wild panda population.

The smaller Red Panda shares a similar distribution to the Giant Panda but extends further west along the southern forested slopes of the Himalayas as far as into east Nepal. Like the Giant Panda it eats bamboo and like the Giant Panda it has a flattened face, broad mouth and hands designed for holding the bamboo stems.

Unlike the Giant Panda, the Red Panda has a long nine-banded tail, is not much bigger than a cat and more arboreal. They sometimes eat quite a lot of fruit and berries as well as bamboo.

Because they are rather short and the bamboo is tall they stand up a lot to feed. They also like to use the support of a fallen log to walk along whilst they can sample the sweeter leaves at the top of the bamboo plants.

Zoologists argue about the position of both pandas among the carnivore families. They were formerly classed together in the same family of their own but DNA analysis seems to reveal a surprising relationship. The Giant Panda is a bear and the Red Panda is a raccoon.

Two Pandas up a tree in Wolong nature reserve.
卧龙自然保护区内两只大熊猫在一棵树上。

第三章
中国特有的野生动物

熊猫

Pandas feel safe and often enjoy a sunny rest in the treetops, but they are clumsy climbers and sometimes tumble to the ground. No harm done they are tough compact creatures.

大熊猫在树梢上感到安全，往往喜爱在阳光灿烂时在树梢休息，但它们攀爬笨拙，有时会跌倒地上。不过没事，它们还蛮强硬，体格不弱。

中国最有名的野生动物当数大熊猫。大熊猫分布在四川盆地，从陕西省的秦岭山脉到甘肃南部的白水江地区和四川西部的岷山、邛崃山和凉山山脉。在史前时期，甚至历史时期，大熊猫就活动于中国的许多地区。但是气候变化和人类活动的压力迫使它们退到这些避难所。甚至可以说，它们已经退到了墙脚下，这面墙就是巍峨的青藏高原。但正是这高海拔高原将季风水分拦截在川西的峡谷里，这就是这个地区如此潮湿和多云的原因，也是竹丛在亚高山林中生长如此茂盛的原因。熊猫以竹子为主食，几乎不吃其他食物。竹子大面积的开花习性曾威胁到熊猫的生存，竹子开花后，成年植株死亡，新生的实生苗要十几年后才能供大熊猫取食。但是，熊猫正是在这种模式中逐渐演化。

竹子开花会导致一些熊猫挨饿或死亡，但也会驱使一些熊猫向别处迁移。它们穿过峡谷，翻过高山，寻找有未开花竹子的山谷。这个扩散的阶段一方面有利于淘汰弱小和不健康的熊猫，实现自然选择；另一方面也有利于远亲繁殖，保证熊猫种群的遗传健康。

熊猫会穿过农地，但不是到村庄寻求食物，而是要到村庄的另一边去寻找竹子。除了在个别极端情况下，这些动物有时会被误解为需要援助。多数情况下，"抢救"回来的熊猫其实十分健康，应让它们继续漫游，或将它们喂饱，然后放回野外。

中国过于沉迷于人工饲养，目前大熊猫的人工种群已经超过200只。尽管繁殖的成功率提高了，人工饲养的熊猫每年也能生产30~40只熊猫仔，但是依然未能将它们成功地返回大自然。目前，圈养的大熊猫已经能自我维持，不再对野生种群形成压力。应继续保护野生熊猫种群，需要设计一个自然保护区系统，为熊猫提供安全的栖息地，并为熊猫种群间迁徙提供足够的走廊。近几年来，熊猫保护区的数量翻倍了，目前已经有超过60%的熊猫栖息纳入保护区范围之内。

熊猫也受法律的严格保护，猎杀或出售熊猫者最高将被判处死刑。几年前，有些商人因私藏熊猫毛皮而被判处死刑。即使如此，许多熊猫也不能幸免于难，有许多熊猫因落入猎人为捕捉麝或其他野生动物而设置的猎套而死亡。但总的来说，野生熊猫的数量是在不断增加，已达1600只。但是2008年的大地震将会对熊猫栖息地带来一定的影响，需要调查大地震对野生熊猫种群的影响。

体重较轻的小熊猫与大熊猫的分布相似，但在西面有更广阔的活动空间，沿喜马拉雅山脉南面的有林山坡一直到尼泊尔东部。它像大熊猫一样以竹子为食，也有扁平的脸、宽大的嘴，以及可抓握竹子的前爪。

与大熊猫不同，小熊猫有一条长长的带有九个环的尾巴，体型比猫和多数树栖动物大不了多少。除了竹子外，它们有时也吃水果和浆果。

由于它们相当矮，而竹子很高，它们必须站立起来取食。它们还喜欢爬上倒伏的树干，以便能吃到更甘甜的竹叶。

这两种熊猫在食肉目中的分类地位一直是动物学家争论的一个问题。它们以前被归入一个科，但是DNA分析结果显示，大熊猫属于熊类，而小熊猫属于浣熊类。

The red or lesser panda is a pretty animal with a round flat face, a liking of bamboo and a foxy red coat.

红熊猫或小熊猫是种美丽的动物，脸蛋圆而平，喜吃竹子，红色外皮像狐狸毛一样平滑。

Red panda clambers to reach bamboo tops.

小熊猫笨拙地攀爬到竹林高处。

Bears

There are three true bears in China. The brown bear lives up on the Tibetan plateau and also in the forests of northeast and northwest China. It is a huge bear comparable to the Grizzly of North America. It is the most carnivorous of the Chinese bears and also eats carrion. It still has to take in quite a lot of fruits and roots, or where it can catch fish, crabs and invertebrates. This bear is now very rare.

The Himalayan black bear is much more common. It became endangered due to threat from hunting for meat, fur, bear gall and to protect fruit orchards. Since 1981 it was added to the national protected species lists. Its status has improved in the wild and it is again regarded as a pest in some localities.

It is extensively farmed and several thousand bears are now in breeding farms where bears are raised to produce bear bile which is a valuable Chinese medicine. The bears are fitted with a catheter fitted into the gall bladder. China comes under a lot of criticism from animal right groups over this 'cruel' treatment. But by meeting the needs for bear bile from this stable captive population, each bear can yield bile almost every day for several years, the government have certainly taken a lot of pressure off the wild population which was formerly hunted just to supply a one off yield per bear killed. It is hoped that synthetic bear gall or other substitutes can be accepted to replace demand for bear gall. Meanwhile sanitary regulations are being applied to reduce the cruelty and risk of infection. NGOs have established bear sanctuaries for bears rescued from farms.

The black bear eats a wide range of fruits, insects and meat but especially likes nuts such as acorns and walnuts. In autumn a bear may climb into the crown of an oak tree and make a platform by pulling in all the branches to the centre and stripping them of their acorns. The tree is left in a damaged state and the bear goes off to find another. The tree will recover next year. Another trick is to roll around in a clump of brambles dragging them together into a big ball then easily sitting there picking out the fruit.

The smallest bear in China is the Malayan sun bear. This is a tropical species and only found in the extreme south of Yunnan in Xishuangbanna and in the Huanglianshan hills on the Vietnam border. The sun bear is sometimes called honey bear as it is very fond of honey and will break into the heart of hollow trees if there is a bee's nest there - feeding on not just the honey but the waxy comb, grubs and all.

The sun bear is an agile climber and spends a lot of time up in the tree canopy. Sometimes it will stay up a fruiting strangling fig tree for several days until it has cleared out most of the fruit.

China's smallest bear - the Malayan sunbear - has long sharp claws that enable it to climb trees quickly. Tree trunks bear the scars where the bears have climbed.

中国最小的熊-马来熊-爪长而尖锐，因而能迅速攀登树木。熊攀爬过处都留有伤疤。

熊

在中国有三种真正的熊。棕熊生活在青藏高原以及东北和西北的森林里，体形庞大，堪比北美的灰熊。它是中国熊类中肉食程度最高的，但也取食腐肉、鱼、蟹和无脊椎动物，还吃大量的果实和根茎。目前，棕熊已经十分罕见了。

黑熊更为常见。因为人类要获取熊肉、毛皮、熊胆和保护果园，黑熊常遭到杀戮，现已处于濒危的境地。自从1981年以来，黑熊就被列为国家保护动物。它的野外数量已经增加。在一些地方它被视为害兽。

黑熊被广泛饲养，养殖场内黑熊的数量已达几千头，养殖黑熊的目的是生产熊胆粉，它是一种珍贵的中药材。活熊取胆需要将一根导流管插入胆囊，定时引流。这种方式受到了动物权利组织的严厉批评，认为这样对黑熊太"残忍"。但是，由于圈养黑熊能每天生产胆汁，且能连续生产好几年，而人工养殖的黑熊能提供稳定的熊胆产品，满足人们的需求。以前

采用的是猎熊取胆，每猎杀一头黑熊只能取得一个熊胆，因此，活熊取胆实际上减小了对野生黑熊的威胁。希望人们能很快接受人工合成的熊胆产品，或用其它产品来取代天然熊胆。同时，正在通过卫生条例来改善熊的饲养条件和黑熊感染的风险。非政府组织已经建立了熊的避难所。

黑熊的食物范围很广，包括多种果实、昆虫和肉，尤其喜爱橡子和核桃等坚果。秋天熊会爬到橡树的冠丛中，将周围的树枝拉到中间，建成一个平台，然后取食橡子。橡子吃完后，树木也被损坏了，于是熊又去寻找其他的橡树。被损坏的橡树能在翌年恢复。熊的另一诀窍是在灌木丛中打滚，从而把灌木卷成一团，然后坐下来悠闲地采食上面的果实。

中国最小的熊是马来熊。它属于热带物种，仅见于云南的西双版纳和中越边境的黄连山。马来熊也叫蜜熊，因为它喜爱吃蜂蜜，会闯入空心树中，寻找蜂

巢，它不仅吃蜂蜜，也吃蜂蜡、蜂蛹等。

马来熊攀爬敏捷，多数时间待在树冠上。有时，它会连续好几天待在果实长满枝头的榕树上，直到将果子吃完。

Brown bears are the largest carnivore of the high plateau.

棕熊是高原上最大的食肉动物。

The black bear is getting more numerous since becoming a protected species, despite continuing use in breeding farms to produce bile.

尽管养殖场继续使用黑熊生产胆汁，但自从黑熊成为受保护物种后，它们的数量在增加。

Apes and Monkeys

Golden monkeys

The golden or snubnose monkeys are a very specialized group of leaf eating monkeys confined to China. Another closely related species occurs in the limestone forests across the border in Vietnam. The three Chinese forme were formerly all regarded as races of a single species the Golden monkey. However it is clear from their distribution, genetics and morphology that there are three distinct species in China. All are large, have a saculated stomach that allows them to digest cellulose and other very course materials and all live at high altitudes. They can aggregate into huge troops off many tens or even hundreds of animals. They move over large ranges in the sub-alpine forests. They eat a wide range of plant materials—young leaves, shoots, fruits and particularly the fleshy lichens that grow on the forest trees in the humid high-altitudes. In some months they range high up to the top of the tree line. But in the cold of winter they come down lower into the valleys.

All species have a flat wide face with the upturned nose and in the males lappets on the side of the face, which gives them a unique appearance. The Sichuan Golden monkey is truly golden with a long shaggy coat and strangely blue face. The Sichuan Golden monkey is found from West Sichuan around the top of the Sichuan basin and through to the Qinling mountains of Shaanxi Province and as far east as Shennongjia range in Hubei Province. A much rarer and more limited species the Grey snubnose monkey is confined to Fanjingshan mountain in Guizhou Province. The Yunnan or Black snubnose monkey is found in the mountains of northwest Yunnan and ranging into the southeast corner of Xinjang. Here one can find bands of up to 200 animals living in the conifer and mixed oak forests from 3400m to over 4000 m in altitude.

Family portrait of Sichuan golden monkeys.

川金丝猴的全家合影。

Two young Yunnan snub-nose monkeys feeding.

两只滇金丝猴幼猴在喂食。

猿与猴

金丝猴

　　金丝猴是食叶猴中非常特殊的一类，仅分布在中国。另一紧密相关的物种见于中越边境的石灰岩森林。国内的金丝猴以前被分为三个亚种，然而，它们的分布、遗传学和形态学特征都表明，它们是三个独立的种。它们体型都很大，有一个分室的胃，能消化纤维素和其他十分粗糙的食物，全都生活在高海拔地区。它们能聚集成几十或几百只的群体。它们在亚高山森林里的活动范围很广。它们的食物包括嫩叶、芽和果子，尤其爱吃湿润高海拔区树木上的肉质地衣。每年有几个月，它们在林线以上的高海拔区活动。但是，到了寒冷的冬天，它们会下到山谷活动。

　　所有的金丝猴种的脸都扁平，鼻子上翻。成年雄猴的上唇嘴角处生有一个肉垂，这使它们看起来很独特。四川金丝猴真是金色的，有一身长长的、粗浓的金色毛发，有一张奇特的蓝脸。川金丝猴见于川西、四川盆地北缘向北到陕西省的秦岭，东面到湖北省的神农架。黔金丝猴只分布在贵州省的梵净山，其数量和分布面积十分有限。滇金丝猴见于滇西北至西藏东南隅的山区，在这个地区，有时能看到数量多达200只的猴群生活在海拔3400米到4000多米的针阔混交林中。

Family of Sichuan golden monkeys huddle together for warmth, protecting their babies.

一群川金丝猴挤靠在一起取暖，保护婴幼。

Langurs

Various species of Langur or leaf-eating monkeys occur along the southern slopes of the Himalayas and in the forests of SW China. The most common is the Francois's langur which is an all black leaf monkey with whitish cheeks. The crown of the adult has a pointed peak of black hair. The infant is bright orange when it is born, fading to yellow before it becomes gray and eventually black like the adult. These monkeys live in the forest on limestone hills in Guangxi and Guizhou extending to east Yunnan and across the national borders into both Vietnam and Laos.

A gray leaf monkey with white rings round its eyes occurs in the forests of Yunnan known as Phayre's langur. Another species with bright yellow eyes is the little-known Shortridge's langur.

The white-headed langur is a rare species confined to the limestone forests along the Guangxi Vietnam border. It is a black langur with contrasting white head and white lower half to the tail. Groups sleep in cliff caves and farmers climb up to look for the menstrual blood or dried placentas which are sold as precious medicine.

The female Cao Vit gibbon. Only 200 of these animals remain in the limestone of the Guangxi - Vietnam border. ECBP helps local government establish a nature reserve for this and other rare species.

雌性东部黑冠长臂猿现仅有200只存活在广西－越南边境的石林一带。中国－欧盟生物多样性项目帮助当地政府建立了一个自然保护区来保护这种长臂猿和其他稀有物种。

叶猴

多种叶猴出现于喜马拉雅山南坡和中国西南地区的森林里。最常见的是黑叶猴，它全身黑色，但脸颊白色。成年黑叶猴的头顶有小小的黑色冠毛。幼猴刚出生时呈鲜亮桔黄色，成长过程中逐渐变黄、变灰，直到成年后变成黑色。这些猴子生活在广西和贵州的石灰岩森林里，活动范围延伸到云南东部，跨越国境，直到越南和老挝。

菲氏叶猴出现于云南森林，眼睛周围带有白色环。另一种眼睛明亮，呈黄色，但更不为人知的是戴帽叶猴。

白头叶猴是稀有物种，仅生活在沿广西-越南边境的石灰岩森林。黑叶猴的白色头和半截白尾形成鲜明对比。黑叶猴群居于悬崖上的洞穴中。当地农民常爬入洞中寻找黑叶猴的经血或干燥的胎盘，把它们当作珍贵的药材出售。

Gibbons

Gibbons are the smallest of the apes and lack tails. China has six species of gibbon.

The most westerly form - the White-browed or Hoolock gibbon lives only to the west of the Nujiang River and to the south of the Gaoligong range in western Yunnan, but extends through northeast India and Mynamar. The white-handed gibbons is found further south, also to the west of the Nujiang River and its rage extends through most of Thailand and into Malaysia and Sumatra in Indonesia. Black gibbons have been divided into several species with the Crested Gibbon found in the Ailaoshan and Wuliangshan ranges of Yunnan, the Hainan gibbon found only in the southwest of Hainan Island, white-cheeked gibbons found to the east of the Nujiang in the south of Yunnan and extending into Indochina and a very rare form of black gibbon formerly classed as a race of the Hainan gibbon but now separated as

a distinct species called the Cao Vit gibbon occurs along the Guangxi Vietnam border and maybe only 200 individuals of this species survive.

Gibbons live in much smaller families than most other monkeys—generally an adult female an adult male and one or two children. They are highly territorial and defend their territories with displays include the chorus of very loud singing given most mornings just after dawn. Gibbons are frugivores and eat a large proportion of small fruits, especially figs in their diet.

长臂猿

长臂猿是猿中体型最小的，无尾巴。中国有6种长臂猿。生活在中国最西边的长臂猿是白眉长臂猿，其活动范围为云南省怒江以西和高黎贡山以南的地区，并延伸到印度的东北部和缅甸。白掌长臂猿的分布区更往南，到怒江的西面，包括泰国的大部分地区，并一直延伸到马来西亚和印度尼西亚的苏门答腊。黑长臂猿被分成几个种：黑胸长臂猿见于云南的哀牢山和无量山；海南长臂猿见于海南岛的西南部；白颊长臂猿见于云南怒江的东部，并一直到印度支那。还有一种非常罕见的黑长臂猿，以前被认为是海南长臂猿的一种，但是现在被视为一个独立的种，即东部黑冠长臂猿，出现于广西-越南边境地区，可能仅有 200 只存活。

　　长臂猿的群体比多数其他猴子群体都小，通常由一只成年雌猿、一只成年雄猿和一只或两只幼猿组成。它们地域性很强，会通过各种方式展示自己，捍卫领地，其中一种方式是，在拂晓，集体大声吼叫。长臂猿是食果动物，采食很多小果实，尤其爱吃榕树的果实。

Young rhesus macaques checking out their smart good looks.

年幼的猕猴们检阅着自己时髦的外表。

Male of the white-cheecked gibbon, hanging in a large fig tree.

雄白颊长臂猿，悬挂在一棵巨大的无花果树上。

Macaques

The commonest monkeys in China are the macaques. These are generalized robust monkeys with relatively short tails and usually have rather red faces. The commonest is the Rhesus monkey which is found over most of southern China extending as far north and east as Beijing.

Also widespread is the Tibetan macaque a darker, more chocolate colored macaque ranging from the east of the Tibetan plateau and through southern China to Fujian Province. The Pigtailed macaque is a fierce-looking monkey with a short tail. It occurs only in the southwest Yunnan; and the Stump-tailed and Assam macaques both occur in Yunnan and Guangxi as far as Guangdong.

All these macaques live in multi-male bands so there may be 10 to 30 or even 40 animals traveling together in a single group. Several mothers will group together to form nursery groups to play with and care for their children at the edge of the main group and out-of-the-way of the dominance battles that rage from time to time between the larger males.

Macaques are omnivores and eat everything from fruits to leaves to invertebrates even small vertebrates if they can catch them. They are equally at home in threes as on the ground, whereas all the leaf monkeys and gibbons are almost entirely arboreal. In some localities where tourists feed monkeys such as the sacred mountain Omeishan in Sichuan, or the monkey kingdom of Nanwan in Hainan, the monkeys become over-crowded, overly bold and rather aggressive.

猕猴

在中国，最常见的猴子是猕猴。它们普遍精力充沛，尾巴相对短，且通常有一张相当红的脸。最普通的是猕猴，见于中国南方的大部分地区，往东北方向延伸，最远能到达北京。

藏酋猴的分布也很广泛，但颜色更深，更多的呈棕褐色，分布范围从青藏高原东部，进入华南，到达福建省。豚尾猴面目狰狞，尾巴短，仅见于滇西南；短尾猴和熊猴均见于云南和广西，最远分布到广东。

所有这些猕猴都生活在雄性个体多的群体中，在同一猴群里，可能有10~30只，甚至多达40只在一起活动。几只母猴在一起组成一个保育组，它们排在猴群的边上，保护着幼猴，并在雄猴争夺王位而战时，将幼猴带离战场。

猕猴是杂食动物，它们既吃果实和树叶，也吃无脊椎动物，也不放过能捉到的小型脊椎动物。无论是在树上还是在地面，它们都是活动自如，而几乎所有的叶猴和长臂猿都是树栖动物。在一些游客可以投饲的地方，如峨眉山和海南南湾猴岛，猴群的数量已经过高，猴子也大胆妄为，甚至会伤害游客。

The pig-tailed macaque is a famed raider of maize fields and is tough enough to stand off farmers dogs.

豚尾猴是玉米地里臭名昭著的盗贼，它顽皮不驯，也不怕农场上的狗。

Elephants

Some people are surprised to learn there are still wild elephants in China but there are and they are not so difficult to visit and see. In historical times, Asian elephants roamed over much of southern China but as humans have cleared lands and expanded their agriculture elephants have been pushed back to the frontier areas along the Myanmar and Lao borders of Yunnan Province. Here several hundred of these magnificent beasts remain, although they are not the most comfortable neighbours to the villages that have sprouted up around their habitat.

Mostly the elephants consist of matriarchal family groups of one or two large females with a bevy of younger animals of various sizes. The young males sort sharp white tusks. But wandering alone or occasionally in small bachelor parties are several huge males bearing large ivory tusks.

Elephants are creatures of habit. They make regular travels to ancient salt licks and water holes and if humans come and make clearings, farms villages or highways in the intervening lands they are liable to face herds of elephants wandering through from time to time.

Elephants enjoy eating most of what humans like to grow in their farms and gardens. These are so much sweeter than the coarse foods that can be found in the primeval jungle, so elephants are notorious raiders and do considerable damage to crops and property. They are especially fond of rice, bananas, all kinds of palms, sugar cane, fruits and various other vegetables. They also smell salt in peoples' houses and sometimes break through walls to steal this mineral from household kitchens.

Defending crops and villages is a challenging and sometimes dangerous task. Villagers construct barriers and electric fences. They build platforms in trees and stay up all night to deter elephants from entering their fields. They fire muskets or fireworks or bang drums to scare off marauding elephants and sometimes in desperation shoot elephants to protect their precious property.

Elephants retaliate and sometimes attack farmers or smash houses out of bad temper. It is sometimes a surprise for visitors to discover that elephants can be so dangerous when the cartoons and television shows have always portrayed elephants as gentle friendly animals that can be domesticated and help humans in their labours.

Many people do not realize how dangerous wild elephants are. So tourists visiting the elephant Valley of Sanchahe in Xishuangbanna are usually ill prepared for the risks they are taking as gangs of noisy people shout with glee as elephants emerge from the forest to bathe in the river. People throw food. People

Elephants look tame but they can be very dangerous. Tourists in Elephant Valley must take great care not to get too close, especially if elephants have wounds from farmers gun shots.

大象看似驯服，但也可能非常危险。去大象谷游玩的游客必须非常小心，切不要离大象太近，特别是如果大象有枪伤的话。

go too close to take photographs or to be themselves photographed close to the elephants. And not surprisingly there are accidents. Two people have been killed in the last year.

Reserve officials are faced with claims for the damage done by elephants to farmlands around the nature reserve. They imagine that the forest does not provide enough food for the elephants, and think that is why the elephants need to come out to eat in farmland. In fact forests are large and contain easily enough food for the quite small numbers of elephants remaining in Xishuangbanna. The problem is that elephants are conservative and like to roam over those areas they have used for many years. Even though such areas have now been converted to farmlands, the elephants still feel they are their own territory to use.

It should be recognized that if farmers decide to move into elephant territory and plant sweet rice and sugar cane too close to wild elephants, they must expect elephants to come and raid those crops. It is their responsibility to protect those crops by making ditches or erecting electric fences. It is not the responsibility of the reserve officials to control the natural behaviour of wild elephants. There is a plan to clear forest in the center of the reserve and plant it with sugar cane so that elephants will be attracted inwards to the centers of the reserve. This is a rather crazy idea. It will cause the destruction of the very core of the rainforest reserve which is not just for elephants but for thousands of other species. Moreover it probably will not work. The elephants routes are determined by seasonal patterns, by the distribution of the minerals. They will still raid the outer farms.

Three lady elephants stride their stuff through a grassy glade in Xishuangbanna.

三只雌象在西双版纳的草地上悠闲阔步。

Mother elephant with two youngsters. The baby is only a few months old.

母象与两只小象。幼象只有几个月大。

亚洲象

听说中国还有野象,而且看到它们还很容易,有人可能会很吃惊。在历史上亚洲象曾广泛分布在中国南部,但是随着人类毁林造田,扩展农业,以及气候变化,大象被赶到了中国、缅甸和老挝的边境地区。在这个地区,尚存有几百头亚洲象。在大象栖息地周围,随着村庄的扩大,大象常常会给村民们增添麻烦。

多数情况下,大象组成母系群体,由一只或两只大雌象带领一群大小不一的小象。幼雄象有尖锐的獠牙。但是,独自游荡或偶尔成小群一起活动的通常是成年雄象,它们的獠牙又大又长。

大象的生活规律很强,它们会定期去固定的地点舔盐,固定的水源喝水,因此,如果人们在大象经过的线路上开荒种地、造农园,建村庄,甚至修公路,就会时不时与象群碰面。

人类在田地和院子里种植的大部分作物都是大象喜爱的食物,因为它们比原始森林里的食物更加甜美可口,所以大象对周边地区的庄稼和财产造成了极大

的破坏,因而成了恶名昭彰的破坏者。它们特别喜欢吃水稻、香蕉、各种棕榈树、甘蔗、水果和蔬菜。它们也能嗅到人们屋子里的盐,有时会破壁而入,到厨房偷盐吃。

保卫庄稼和村庄极具挑战性,有时也很危险。村民建筑围栏和电子栅栏防止大象入侵,并在树上搭建小屋,整夜在小屋里守候,阻止大象进入田地里。他们或燃放火枪,或点燃鞭炮,或敲锣打鼓来驱赶前来掠食的大象,但当这些手段都不见效时,为了保卫珍贵的财产,有时不得不朝大象开枪。

大象会进行报复,有时发怒时会攻击农民或毁坏农民的住房。电视节目卡通片总是把大象描述为温和、友好的动物,能被驯养,还帮助人们从事体力活,有时,游客会惊讶地发现,大象又是如此地危险。

在中国,对大象的防范意识很低,受电影和卡通片的影响,多数人认为大象不会伤害人,是友好的动物。许多人意识不到野象的危险性。因而,游客在参

观西双版纳三岔河的野象谷时,通常对大象的防范准备不足。人们看到大象从森林出来,到河里洗澡,就大声地欢呼雀跃,并向大象扔食物。也有人走近大象拍照或与大象合影。人们如此靠近大象,出现意外就不足为奇了。去年,就有两人因此丧生。

由于大象破坏了自然保护区附近的庄稼,村民则向保护区的官员索赔。人们想当然地认为,森林给大象的食物不足够,所以大象要跑出来吃庄稼。实际上,西双版纳的大象尚存数目不多,森林很大,有足够的食物供大象生存。问题是大象较为怀旧,喜欢在它们生活多年的土地上漫游。尽管有些土地已经被开垦成农用地,但大象仍感到那是它们的领地。

农民如果在大象的领地内种植甜水稻和甘蔗,应预想到由于离野象太近,大象就会出来破坏庄稼。他们自己有责任挖掘隔离沟或竖立电网保护庄稼,保护区的官员没有责任控制野象的自然行为。有人计划在保护区中心砍伐森林,种植甘蔗,以便将大象吸引到保护区的中心,这是个相当奇怪的想法。这将对这个热带雨林保护区造成破坏,影响到大象和其他数千个物种的生存。此外,这种做法也不见得会起作用,因为大象的行动路径是有季节性的,也和矿物质的分布相关,就是保护区中心种了甘蔗,它们仍然可能会闯入农田。

Sanchahe elephants enjoy the mud together.

三岔河的大象群在烂泥中享乐。

Deer

There are 24 species of deer in China, more than any other country. These vary enormously from the dainty antler-less mouse deer of the tropical forests of Xishuangbanna to the huge moose of northeast and northwest China.

All deer are hunted for their tasty venison. Villagers set snares along animal trails in the forest. All are rarer than they should be but one group in greatest demand are the musk deer. Four species of musk deer are distributed from the Himalayas to the mountains of the northeast and they are especially hunted because the male has a musk gland on its under belly that is highly prized as both a medicine and in the perfume industry. These are difficult deer to breed in farms and it is still the wild animals that supply the trade. Snares cannot distinguish the sex of the prey so many musk-less females are killed by accident.

Four species of barking deer or muntjac occur in the southern forests. These deer have simple antlers and large canine teeth. They get their name from the loud alarm calls they emit when disturbed—a loud explosive bark—rather like a dog. The tufted deer of the mountains of western Sichuan is a close relative of the muntjac's but lacks any antlers at all.

In the south west of China we find the red barking deer, in the rest of southern China a smaller sandy coloured Reeve's muntjac, in the mountains of southeast China the endemic black muntjac and in the Gaoligong mountains on the Yunnan Myanmar border another dark form occurs.

Dainty roe deer occur in the northwest and northeast of the country, together with the larger red deer. Sambar deer range from the Himalayas through central China and some specialist deer such as Chinese water deer, hog deer and the reintroduced Pere David's deer all prefer wet grasslands. The latter is a strange looking deer given the Chinese nickname of si bu xiang (four unlikes) because it has "the tail of a donkey, the head of a horse, the hoofs of a cow, and the antlers of a deer".

Red deer occur in both the northeast and northwest of China, feeding in grasslands but hiding up in forests. A rarer race exists in the Himalayas.

马鹿在中国东北部和西北部都有，它们觅食于草原，但藏身在森林。喜马拉雅山上有一种更罕见的物种。

鹿

中国有 24 种鹿，比其他任何国家都多。包括从西双版纳热带森林的麂鹿到北方的驼鹿，它们的体形大小不一，差异极大。

所有的鹿都因其美味的肉而被猎捕，村民们在林中沿着鹿的踪迹设置猎套，因此上，所有的鹿都比它们应有的数量要少。人们对麝的需求最大，从东北的森林到喜马拉雅山脉，分布着四种麝。由于雄麝的脐下有一个麝香腺，而麝香是一种名贵的药材和香料，因此，麝面临的猎捕压力特别高。麝的人工饲养很困难，因此目前麝香的供应依然来自野外。由于猎套并不能分辨猎物的性别，许多产麝香较少的母麝也被误杀。

在南方的森林里生活着四种麂子。麂子有简单的角和突出的犬齿。受到惊扰时，麂子会发出急促的象狗吠一样的叫声，因此也被称为吠鹿。四川西部山区的毛冠鹿是麂子的近亲，但没有角。

在西南地区分布着赤麂，在南方的其他地区分布着体型更小的草黄色的小麂；在西南山区有黑麂，在中 - 缅边境的高黎贡山分布着贡山麂。

体型纤细的狍子分布在西北和东北地区，和它同域分布的是体型大得多的马鹿，水鹿的分布范围包括从喜马拉雅山脉到华中的大部分地区。中国还有几种对栖息地要求比较严格的鹿，如獐、豚鹿和重新引入的麋鹿，它们都喜欢在湿草地上活动。

后者因外貌独特而被叫做"四不像"，它有驴尾、马头、牛蹄及鹿角。

Red muntjak with antlers in velvet.

赤鹿天鹅绒般的茸角。

A group of female Eld's deer look alert in the Datian nature reserve of Hainan.

海南大田自然保护区，一组雌坡鹿警觉着。

Eld's deer are found only on Hainan. Other races occur in northeast India and Indochina. The male antlers sweep forward giving the species the alternate name brow-antlered deer. These deer now only survive in large fenced reserves, but adequate grassland remains on Hainan to consider releases if hunting can be controlled.

坡鹿仅见于海南。其他品种见于印度东北部及中南半岛。雄鹿角非常宽。这些鹿现在只生存于保护区围栏内，但如果狩猎能被控制，海南仍有充足的草地，可考虑放生野外。

Wild Horses

There are three species of wild horse in China—wild nervous creatures of the arid steppes and deserts. What a sight to see a large herd galloping through a cloud of dust; small colts hurrying to keep up with their mothers and young stallions showing off their power and speed.

On the high plateau we find the kiang—a wild ass that lives in large herds. The animals range up to 5000m and feed mostly on grasses. The kiang has white underparts but a brown back and shoulders with a thick dark down the back and onto the long bushy tail.

The Kulan or wild ass of the northern deserts lives at much lower altitudes. The tail is fluffy only at the tip and the upperparts are a more sandy colour than the slightly smaller kiang. Kulan are able to go long periods without water and can supplement their diet with many desert shrubs when grass is scarce.

The third species of horse is a single subspecies of the ancestor of Asiatic domestic horses - the Przevalskiís horse. The species bas been crossed with domestic stock so many times that it is regarded as extinct in the wild. However, 12 founder animals that had been taken into captivity were used to build up a new population. There were some problems with impure gene lines but finally a rather pure form has been raised and reintroduced back to the wild in the Kalamaili Nature Reserve near Dunhuang in the Gansu corridor. About 35 animals now run free there.

Przevalski's horse is a sturdy animal with fawn brown coat, black socks and dense black tail. It has a thin dark stripe down the back and a stiff upright mane. It remains the model for the famous clay horses of the Tang Dynasty and early Chinese paintings of horses and the mounts of Chinese and Mongolian horsemen.

Przewalski's horse. Captive bred and ready for reintroduction.

普氏野马由圈养孕育，准备放生以恢复野生马群。

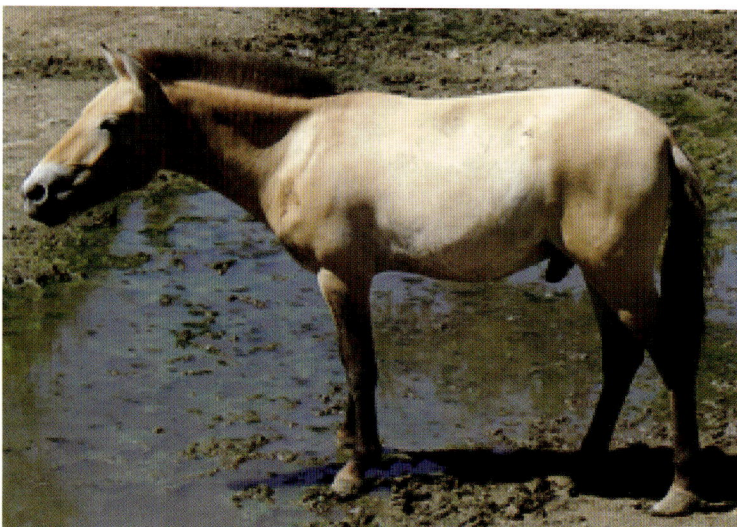

野马

中国有三种野马，它们居于干旱的大草原和沙漠，是最令人兴奋的野生动物。大群的野马飞奔，尘土飞扬；小马驹紧跟在母马身后，年轻的牡马在前面领跑，展示着它们的力量与速度。

在高原上，有一种群居性的野驴——藏野驴，它能生长在海拔高达 5000 米的地区，主要以草为食。藏野驴下体白色，背和肩棕色，背底部的深黑色一直延伸到浓密的长尾。

北方沙漠上的蒙古野驴生活在海拔低得多的地区，它只在尾尖上长着一簇长毛，上体比体型略小的藏野驴呈更深沙色。蒙古野驴能在缺水的情况下，长途跋涉，在缺少草料时，能食用多种灌木。

第三种野马——普氏野马——是亚洲驯养马的祖先。由于已经与驯养马杂交多次，人们认为这个物种在野外已经灭绝。人们把人工饲养下的 12 只野马放在一起，试图建立一个新种群。一开始马的血系有些问题，但最终人们培育出了一种纯血系的野马，并将它重新引进了位于河西走廊敦煌附近的卡拉麦里自然保护区。目前，野外大约有 35 只野马。

普氏野马很健壮，毛色为浅黄褐，腿黑色，尾巴深黑。其背下有细小的黑条纹和僵硬直立的鬃毛。著名的唐三彩、早期的中国绘画和蒙古骑士的坐骑都是以它为模型。

A group of kiang enjoy the warm weather and good grazing in the short summer of Changtang nature reserve.

羌塘自然保护区短暂的夏季，一群藏野驴正享受着温暖的天气和牧地上的美食。

Goat Antelopes

One off the strangest looking animals in China is the takin. Takin is a goat-antelope - a member of a group of animals half away between the goats and antelope. The takin is the largest member of this group and is a shaggy cow-sized creature with a face and horns reminiscent of an African wildebeest. Takins live in small family groups but gather into larger herds during the summer when they range high up on the mountains and emerge above the tree line to feed in the alpine meadows. For most of the year they live in within the forest where they eat grass and browse herbs and bushes in forest clearings.

They have a strong liking for minerals and have developed mineral licks and salt springs which they visit regularly. The takin's liking for minerals will is so strong that they sometimes walk into villages or abandoned farmhouses and lick and eat cement and mud daub walls to obtain salts and other minerals.

Takins from the Qinling mountains of Shaanxi Province are a brighter, more golden yellow color than those in Sichuan and Yunnan. In all areas, takin have been heavily hunted and their numbers are quite reduced. But in those reserves where they are well protected their numbers are now increasing again and you can see quite large herds of these extraordinary animals. They can be rather aggressive animals and there have been odd cases of hunters being charged or even killed by them.

Other goat antelopes in the middle mountains of China include the serow and the smaller gorals. Gorals are dainty creatures and four species inhabit different regions of China. They may be found on very steep mountain slopes all the way from the Himalayas in the West to the forests of Changbaishan in northeast China.

Serow extends southwards all the way to Indonesia. They are medium-sized goat-like animals with a shaggy and blackish coat. Their horns are recurved and heavily annulated. They live especially in the rockiest limestone areas of southwest China.

Wild serow resting in a rocky streambed.

野生鬣羚在多岩石的河床上小憩 。

山羊羚

在中国,山羊羚是长相最奇特的动物。山羊羚是介于山羊和羚羊之间的一种动物。山羊羚是同类动物中体型最大的,大小如奶牛,一身皮毛蓬松,面部和角类似非洲角马。山羊羚成小群生活,但在夏天会聚集成大群,到林线以上的高山草甸上觅食。一年中的大部分时间里,它们都生活在森林里,以林间空地上的草和灌木为食。

它们非常喜欢矿物质,经常舔食矿物质和盐泉,它们有时甚至会到村庄或废弃的农舍去舔食水泥和泥墙。

陕西省秦岭的山羊羚比四川和云南的山羊羚色彩更亮,更黄。在过去,各地的山羊羚都被严重捕猎,数量大减。随着保护措施的落实,山羊羚的数量开始回升,人们能看到大群这种奇特的动物,在有些受到良好保护的保护区内,它们的数量已经出现过高的迹象。山羊羚好斗,攻击性很强,特别是离群独自行动的"孤羊",危险性更大。近年来山羊羚伤人的事故屡有发生。

其他羚羊包括鬣羚和体型更小点的斑羚。斑羚是种很优美的动物。中国有四种斑羚,见于不同的地区。从西部喜马拉雅山脉到东北的长白山,在林中陡峭山坡上都能见到它们的身影。鬣羚的生活范围往南一直延伸到马来西亚。鬣羚是一种体型中等,类似山羊的动物,一身毛发蓬松、略带黑色。它们的角向内弯,有鲜明的环状纹。它们尤其喜欢生活在中国西南部多石的石灰岩地区。

Himalayan goral live on open slopes of the high mountains.

中华斑羚生活在高山区无覆盖的山坡上。

The horns of takin are dangerously sharp and hunters have been killed by an upwards swipe from a bull takin.

羚牛的茸角锋利危险,曾有猎人被羚牛角刺死。

Family of Qinling's ' golden' takins. The calf is dark as in other races.

金色的秦岭羚牛一家子。其他种类的幼畜色彩暗淡。

Golden takin in the Qinling Mountains of Shaanxi are a paler straw colour than takin in other areas.

陕西省秦岭山区的秦岭羚牛比其它的羚牛色泽较淡，呈淡秸秆色。

Wild Cattle and Sheep

Wild cattle are rather limited in China. There are records of banteng in the south of Yunnan but no banteng survive today. More common is the gaur, which are the largest wild cattle in Asia. A few remain in the forests of Xishuangbanna in the southwest of Yunnan. Other than those tropical species of wild cattle there are only the wild yaks that are still found in remote areas of the Tibetan plateaut. Formerly wild yaks were abundant but they have been consistently captured and hunted or crossbred with cattle to give rise to the now large herds of domestic yaks that are raised throughout the former range of the wild yaks.

Blue sheep graze in large herds on the sparse grass of high alpine meadows of the Tibetan plateau and great mountain ranges of western China. They are among the favourite prey of the snow leopard and wolf so remain ever watchful and can run swiftly over the steep terrain.

The largest and most impressive wild sheep in China is the rare argali or Marco Polo sheep. The rams of these sturdy animals have massive recurves horns which they crash together in the mating displays as rivals challenge for the right to form harems. At this time the sound of banging horns echoes for miles through the cold air of the western mountains where they live.

Wild Yak stands defiant

野牦牛傲立牛群。

野牛和野羊

　　野牛在中国的分布相当有限。在云南南部曾发现过白臀野牛，但是现在已经不复存在。更普通的是白肢野牛，它是亚洲最大的野牛，在西双版纳还有少量存活。除了热带野牛外，在青藏高原的僻远地区还发现有野牦牛。以前，野牦牛的数量很大，但是它们不断地被捕捉和猎杀，而且与黄牛进行杂交培育出犏牛，现在犏牛已经遍布野牦牛以前的活动范围。

　　岩羊以大群生活在青藏高原及华西高山区的高寒草甸稀疏的草地上。它们是雪豹和狼爱猎食的食物，因此它们时刻警觉着，还能在悬崖峭壁上健步如飞。

　　中国体型最大的给人印象最深刻的野羊是盘羊。盘羊身材强壮，巨大的角向后向外螺旋形伸展，为争夺配偶时，雄羊低头向对方猛冲，羊角抵撞的轰鸣回荡在西部山区寒冷的空中，在数公里外都能听到。

Ram argali in fighting mood during mating season.

盘羊交配季节好战爱斗。

The gaur roams the forests of SW China in Yunnan and SE Xijang.

印度野牛在中国滇西南的和藏东南的森林常见。

Gazelles and Antelopes

Many people are surprised to learn that there are gazelles and antelopes in China since these are creatures more associated with the African bush; but in fact there are six species grazing on Chinese grasslands and not so many years ago, huge herds of hundreds of thousands of gazelles would migrate annually between China and Mongolia and gazelles were formerly found on the plains just outside Beijing.

The two antelope species are highly endangered. Saiga occur in Russia, Mongolia, Kazakstan and northern China but they have been so persecuted for the supposed medicinal properties of their horns, they became extinct in China and nearly so in Russia. Various attempts have been made to reintroduce a colony in NW China and currently there are about 100 animals surviving there.

The Tibetan antelope or Chiru is more numerous but also much reduced from its former numbers. The problem here is the fine shatoosh wool of the chiru. This wool is made to weave the softest and finest scarves in Afghanistan and Northern India and fetches a high price. Poaching has decimated the wild flocks but several thousand still roam inside the large protected areas of Changtang, Arjinshan and Sanjiang on the Tibetan Plateau.

Goitered, Mongolian and Tibetan gazelles strut their way across the grasslands of their respective regions; but the rarest gazelle—Prezewalski's is critically endangered in two locations around Lake Qinghai and requires concerted conservation activity to ensure its survival.

All the antelopes and gazelles were formerly important in controlling the grass growth and diversity on the vast natural grasslands of northern China and the Tibetan plateau and they in turn were controlled by food availability, harsh environment and the preying of wolves, brown bear and occasionally snow leopards.

Dainty Tibetan gazelle in summer coat.

娇俏的藏原羚穿着夏衣。

原羚和羚羊

获悉中国有原羚和羚羊,许多人很吃惊,因为在人们的心目中这些动物都生活在非洲稀树草原上,但实际上,中国的草原上有6种原羚。前几年,有成千上万的羚羊每年在中国和蒙古之间迁徙。以前,在北京外的平原上就曾经发现过羚羊。

中国有两种羚羊处于高度濒危状况。高鼻羚羊见于俄罗斯、哈萨克斯坦、蒙古和中国北方,它们的角就是名贵的中药材——羚羊角,因此遭到人们的大量捕杀。目前,高鼻羚羊在中国已经灭绝,在俄罗斯也近灭绝。80年代末,中国开始从国外引进高鼻羚羊,在甘肃和新疆进行野放的前期试验工作,目前在半人工状态下饲养的高鼻羚羊有100余头。

藏羚羊的数量更多,但同以前相比,已经大大地减少了。给藏羚羊带来杀身之祸的是它珍贵的绒毛。在阿富汗和印度北部,人们用藏羚羊的绒来毛生产围巾,在国际市场上售价昂贵。偷猎者已经杀害了大批的野生藏羚羊,但是在青藏高原的羌塘、阿尔金山和三江源这些大保护区仍有几千头存活。

鹅喉羚、黄羊和藏原羚昂首阔步地在草原上驰骋,但是最罕见的普氏原羚只发现于青海湖附近的两个地点,已经处于极危状况,需要采取保护措施。

以前,在华北和青藏高原,羚羊和原羚一起在控制草的生长和增加天然草地的多样性方面起着重要作用。反过来,它们也因食物的供应、恶劣的环境以及狼、棕熊和雪豹的捕食而受到控制。

Family herd of chiru feeding in early winter

一群家养的藏羚群在初冬进食。

Rarest of China's antelopes - the male of Przewalski's gazelle in Qinghai.

中国最罕见藏羚羊 - 青海的普氏原羚。

The Last Big Carnivores

Sabre-toothed tigers roamed much of China a mere 15,000 years ago. Today only their fossil jaws remain but tigers, leopards and packs of lean wolves can still be found in the most remote corners. Wolves have never been much appreciated by Chinese society. They were always regarded as dangerous and deadly. They were killed as pests as recently as the wildlife act of 1986. But tigers, though far more dangerous were always regarded with respect and tigers have been revered and used as objects of artistic reproduction throughout the ages.

Four races of tigers survive in China, but their numbers are very small and their status critically endangered. Bengal tigers roam on the southern facies of the Himalayas and range into SE Tibet and extreme west Yunnan. The Indochina tiger occasionally wanders into the forests of south Yunnan from neighbouring Myanmar, Laos and Vietnam. A handful of the endemic South China tiger may remain in the rugged mountains of southern China and a few of the huge Siberian tigers cross from Russia to prey on deer, wild pigs and domestic cattle in the Tumen River area of Jilin Province in NE China.

Far more tigers exist in zoos and breeding farms in China than remain in the wild. Zoos were content to show tigers to the public but the breeding farms were set up to commercially farm tigers for meat and their parts which are regarded as powerful medicine. Since China joined CITES in 1981, this trade has been banned. Tiger farmers could no longer afford to feed their huge carnivores with quality meat. Most such farms have now been converted into tourist attractions.

The problems of tigers are shared by other large predators throughout China. All face persecution by man combined with low densities of prey. The Amur leopard is the rarest of several subspecies. The snow leopard remains a rare creature of mystery in the mountains of Tibet and Xinjiang. Wolves have vanished from most of their former haunts and even smaller carnivores like lynx, golden cat, clouded leopard and soft furred sable are all listed as endangered.

The Siberian tiger is the largest race of tiger and can rip down a cow or red deer with ease.

东北虎是世界上最大的虎，能轻易地捕杀牛或马鹿。

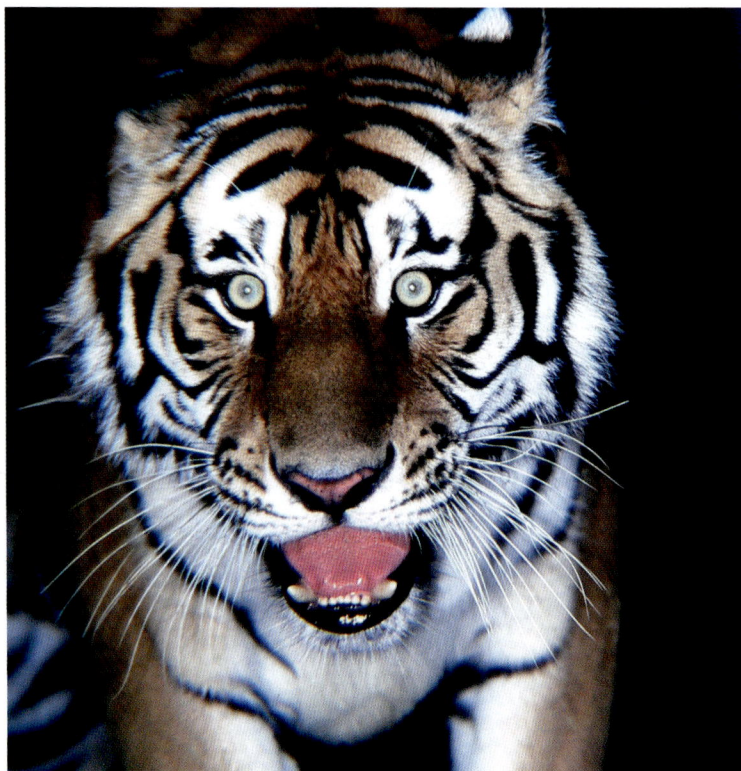

最后的大型食肉动物

15,000 年前，剑齿虎漫游于中国的大部分地区。如今，人们能见到的，只是它的化石了。但是，在许多偏远角落，仍能发现虎、豹和成群的狼。在中国，狼不受人欢迎，人们通常认为狼是危险动物，会危害人的生命。1988 年颁布的《野生动物保护法》仍将狼视为有害动物。尽管老虎远比狼危险，但老虎却受到人们敬畏，并一直是艺术创作的对象。

中国现存的虎有四个亚种，但是数量很少，且处于极危状况。孟加拉虎生活在喜马拉雅山脉的南面，也会进入西藏的东南部和滇西。印支虎偶尔会从邻近的缅甸、老挝和越南进入云南南部。有些华南虎可能还存活于华南崎岖的山林里。少量大型的东北虎从俄罗斯进入东北吉林省的图门江地区捕食鹿、野猪和家牛。

在中国，动物园和养殖场里的虎比野生老虎数量大得多。动物园里的老虎很受游客欢迎，但是养殖场养虎则是出于商业目的，主要生产虎肉和其他器官，人们认为这些器官有壮阳作用。自1981 年中国加入《濒危野生动植物国际贸易公约》以来，对虎产品贸易的控制越来越严格，1993 年中国政府宣布全面禁止虎骨贸易。

老虎养殖者再也养不起这种大型食肉动物。大多数养殖场已改建成了旅游景点。

老虎也面临着其他食肉动物所面临的问题，即被人类猎杀和食物匮乏。在金钱豹的几个亚种中，以远东豹最为稀少。雪豹极为稀有，在西藏和新疆山区神秘出没。狼几乎已经从它们以前的分布区消失了，甚至更小型的食肉动物如猞猁、金猫、云豹及皮毛柔软的紫貂都被列为濒临灭绝。

The handsome snow leopards prey on the herds of Blue Sheep on the steep mountains of the Himalayas and NW China.

在喜马拉雅山和华西北陡峭的山岭上，凶猛的雪豹在岩羊群中猎食。

Yangtze Specials (alligator, baiji, finless porpoise, sturgeon, paddlefish)

Yangtze Alligator hides in the shadows.

藏匿在阴影之中的扬子鳄。

China's longest river—the Changjiang or Yangtze—winds a sinuous route from the high plateau of Qinghai through Yunnan and Sichuan, the great gorges of Central China to the lakes and wetlands of the lower valley in Hubei, Anhui and Jiangsu. It is here in the lower reaches of the Yangtze that forms the home of some special aquatic creatures.

Here we find the rare Yangtze dolphin or baiji, a less rare finless porpoise, Asia's only alligators, a huge sturgeon fish that can grow to 3 metres in length and a curious paddle fish, whose nearest living relative lives in the Missisippi River of the Unites States.

Sadly all are endangered by the pollution and siltation of this huge river and by the fact that the many lakes that used to interconnect with the river and become restocked and replenished each summer flood, are now impounded and isolated by high dykes and flood walls, preventing the easy movement of these large aquatic creatures.

The Institute of Hydrobiology in Wuhan, Hubei have been successful in captive rearing of sturgeon and now release many thousands of young fish into the river each year. The finless porpoise still lives in the Dongting and Poyang lakes and the alligator is also being saved by artificial breeding programmes. But no-one has yet found a way to rear the paddlefish in captivity. We do not even understand its natural diet.

Finless porpoise are still seen regularly in Dongting Lake but the situation of the Chinese river dolphin or baiji is the most desperate of all. It is possible there are none left or only one remaining individual.

The baiji, only living in the Yangtze River, have existed for over 20 million years, but were claimed to have gone extinct on the 8th August 2007.

Between November 6—13 December, 2006, 40 scientists from China, the United States, Britain, Japan, Germany and Switzerland and four other countries searched nearly 3,400 kms back and forth along the middle and lower reaches of international survey intended to find two kinds of freshwater mammals- finless porpoises and Yangtze River dolphins, one of the most endangered mammals. However, to their disappointment, the number of finless porpoises dramatically dropped, besides, there was no slight trace of the Yangtze River dolphins. From 1997 to 1999, the Ministry of Agriculture had organized large-scale monitoring of dolphins, and had found respectively 13, 4, and 4 Yangtze dolphins; but this time, the findings of this survey were zero.

A survey done in the mid-1980s showed that the total number of Yangtze River dolphins had dropped to below 200, of which 50% were in the upper reaches of the River from Shishou County Hubei to Wuhan. They mainly lived in the backwater at the bends of the River or branching streams.

November 4-10, 1997, more than 200 fishery research workers from Hubei, Hunan, Anhui, Jiangxi, Jiangsu, Shanghai and other 6 provinces, cities involved a survey called "Action Plan for Yangtze river dolphins along the middle and lower reaches", organized by the Ministry of Agriculture. This survey, the largest one ever, covered over 1,600 km along the Yangtze River from Zhicheng Hubei to the mouth of the Yangtze River in Shanghai. The 7-day observation found that Yangtze River dolphins were less than 100. Their distribution area also greatly shrunk, dolphins could now only be found in the river between Zhichen and Nanjing and not in Dongting Lake or Poyang Lake. Over the decade, the number of dolphins declined sharply, shockingly less than 100. The experts revealed that the reason of reduction is increasing pollution of the Yangtze River, which has threatened the fish in the River, so the dolphins found it difficult to find enough food to sustain themselves.

The finless porpoise still survives in the Yangtze river and some lakes.

江豚尚存于长江和一些湖泊。

Captive bred alligator ready for release in the wild.

人工繁殖的扬子鳄即将被释放到野外。

长江特有动物
（扬子鳄、白暨豚、江豚、中华鲟、白鲟）

长江是中国最长的河流，起源于青藏高原，蜿蜒流经云南、四川和重庆，穿过三峡进入湖北、安徽和江苏。就是在长江下游生活着一些特有水生动物，包括不久前被宣布灭绝的白暨豚，以及数量急剧下降的江豚；亚洲唯一的短吻鳄——扬子鳄；能长到3米长的中华鲟；以及令人好奇的白鲟，与白鲟亲缘关系最近的物种生活于美国的密西西比河。

不幸的是，由于江水的污染和泥沙，这些动物目前都处于濒危状态。另外，许多与长江相连，能对夏季水灾起缓冲作用的湖泊，如今被高坝和防洪堤坝阻拦和隔离，阻止了这些大型水生动物的自由游动。

湖北武汉水生生物研究所已经成功地人工养殖了中华鲟，如今每年都释放好几千尾幼苗到长江里。江豚依然生活在洞庭湖和鄱阳湖，扬子鳄通过人工繁殖项目获救了。但是，还没有人找到人工养殖白鲟的方法。我们甚至不了解它的天然食物。江豚仍然常见于洞庭湖，但白暨豚的状况可能更令人绝望，可能已经没有剩余的个体了。

已在地球上生存了两千多万年，只在中国长江出没的白鳍豚，于2007年8月8日正式宣告绝种。

2006年11月6日~12月13日，来自中国、美国、英国、日本、德国和瑞士等六国近40名科学家往返近3400公里，对宜昌－上海长江中下游的干流1700公里江段进行了为期38天的长江淡水豚类考察。这一迄今最大规模的国际考察活动，旨在找寻长江孕育的两种淡水哺乳动物——江豚及最濒危哺乳动物白暨豚。然而，令科考人员失望的是，不但江豚的数量大量减少，整个科考过程也未发现一头白暨豚的踪迹。1997年到1999年，农业部曾连续3年组织过对白暨豚进行大规模的监测行动，三年找到的白暨豚分别是13头、4头、4头。此次考察的结果则是0头。

20世纪80年代中期调查时，长江中的白暨豚总头数已下降到200头以下，其中50%分布在湖北省石首县至武汉市上游江段，主要栖息在弯曲河段或弯曲分汊河段的大回水区中。

1997年11月4日~10日，由农业部组织，来自湖北、湖南、安徽、江西、江苏、上海等6省市的200多名科研、渔政工作者，分别在上起湖北枝城、下至上海长江口，全长1600多千米的长江干流上，实施了我国建国以来规模最大的"长江中下游调查白暨豚行动计划"。经过7天的辛勤观测表明，白暨豚现存数量不容乐观，已不到100头。分布范围也大大缩小，枝城以上江段、南京以下江段、洞庭湖和鄱阳湖内，已难以见到白暨豚的身影。十多年的时间里，白暨豚的数量锐减近100头，不能不引起人们的震惊。专家们分析，使白暨豚锐减的另一个主要原因是，长江水体污染日趋严重，鱼类资源迅速减少，使白暨豚赖以生存的食物资源愈来愈匮乏。

Yangtze sturgeon is a huge fish of central China now heavily endangered in the wild but maintained by captive breeding and releases.

体型大的长江鲟栖息华中，野外现已严重濒危，只因圈养繁殖和放生才未绝种。

Pikas and Zokors

The grasslands of China are the home to the last remaining herds of some large ungulates such as gazelles, wild horses, wild sheep and even on the high plateau a few wild yaks. But perhaps the most abundant mammals and ones that play a pivotal key-stone role in the ecology of grasslands are little pikas.

People can be forgiven for thinking pikas are rodents for they have rounded mouse-like ears, live in holes and have rather rat-like or vole-like faces. But in fact the pikas are members of the rabbit family and live in complex social warrens with a maze of underground tunnels and many escape holes.

They are alert little creatures and will dash back down their burrows at the slightest sign of danger. They have to be quick because they are a favourite prey on many a predator's menu. They get eaten by hawks and falcons, chased by foxes and wolves, dug up by bears, pounced on by wild cats, stoats and weasels and trapped down their own burrows by hungry snakes.

Pikas make up for their losses by eating the commonest plants and by breeding like proverbial rabbits. Farmers and herdsmen complain they are too numerous and the Ministry of Agriculture lists them as harmful pests. The government spends a lot of money and effort in trying to eradicate these creatures by spraying millions of hectares every year with rodenticide poison.

This is sad not only for the millions of pikas that are killed by such programmes but because the poison indiscriminately kills a large number of collateral non-target species at the same time. The eradication programme is even more tragic because it may be completely misguided or unnecessary.

Biologists studying the ecology of pikas have concluded that far from being pests and responsible for the overgrazing of productive grasslands, pikas are in fact beneficial animals that serve critical roles in maintaining healthy grasslands and promoting high biodiversity.

Because they are small, pikas prefer the very short, closely grazed grasslands. They remain at quite low density if the grass is too dense or tall. They are most visible and most abundant in overgrazed areas but they are the result of rather than the cause of that overgrazing. It is the human grazing and their domestic animals that have overgrazed the grasslands, but it is too easy to point a finger at the dashing pikas and their maze of tunnels and blame them for the poor pasture condition.

In fact because pikas eat several plants that are noxious or poisonous to cattle, they improve the pasture. Without the pikas and in conditions where the good grasses are being overgrazed it is the obnoxious plants that thrive and start to take over the grasslands.

The northeast pika stands firm sentinel, emitting loud territorial whistles, undeterred by the cold frost.

东北鼠兔立定放哨，哨声响亮，不怕霜冻，保护领土。

Another benefit provided by the pikas is that they both aerate and improve drainage of the grasslands so that what little rain or snow falls, this can penetrate into the ground and nourish the grass, rather than run off the panned ground surface or become evaporated in the dry wind.

The burrows of pikas serve another role as the shelters and breeding sites for several other grassland species. Owls, ground jays, snakes and many other animals use these burrows. Even the little corsac foxes enlarge the original pika burrows for their own dens.

Different pika species are found in the different grassy and alpine regions of China and some are rare or endangered. Yet all are labeled and persecuted as pests unfairly. There is a need to call for an end to this pointless slaughter and tackle in stead the more difficult issue of organizing the correct levels of domestic animal grazing to get maximum sustained yield off Chinas extensive grasslands.

Mice, rabbits and the grassland zokors are also targets of this eradication. The difference for zokors is that they live underground in their burrows, therefore are more difficult to kill. It is hard to believe that the zokors are considered a suitable substitute for tiger bone medicine, so have high economic value. Zokors are now being commercially reared in grasslands as a replacement of the almost extinct tiger resources. Zokor medicinal value may help the species survive. Being a rodent, the zokor, is not so threatened by large-scale poisoning. As a result of its living habits and strong reproductive capacity to adapt to the environment, it is fully capable to deal with this new form of hunting.

The plateau pika in reddish summer pelage. In winter, when conditions get much harsher, the hair is longer and more yellow.

高原鼠兔夏季皮毛略红。冬季气候酷寒时，其皮毛会更长，黄色更多。

Run pika run! Everyone is after you.

跑！鼠兔，跑！每个人都想抓你。

鼠兔和鼢鼠

中国的草地是一些最后剩余的大型有蹄类动物（如原羚、野马和野羊）的家园，在高原地区，甚至也是野牦牛的家园。但是，数量最多而且在草地生态中起关键作用的哺乳类是不起眼的鼠兔。

若人们把鼠兔当作啮齿动物，那是情有可原的，因为鼠兔的耳朵呈圆形，像老鼠的耳朵，生活在洞穴中，脸形酷似老鼠或野鼠。但是，鼠兔实际上是兔子家族的成员，过着复杂的群居生活，鼠兔的居住地中有错综复杂的地下通道和许多逃生洞穴。

鼠兔警觉性很高，若发现丝毫的危险，就迅速蹿回洞穴。它们的速度必须很快，因为它们是许多捕食者所喜爱的食物。它们要逃避鹰和隼的捕食，狐狸和狼的追逐，野猫、鼬类的袭击，饥饿的蛇甚至会入侵它们的洞穴。

鼠兔取食最常见的植物，通过大量繁殖来弥补它们数量上的损失。农民和牧人抱怨它们的数量太多，农业部将它们列为有害动物。政府每年在数以百万公顷的草原上施用灭鼠剂，试图消灭鼠害，花费的人力和财力难以计数。这样的灭鼠工作不但毒杀了几百万只鼠兔，同时还殃及大量相关非目标物种。令人悲哀的是，灭鼠工作很可能是被误导的，或根本就不必要。

研究鼠兔生态学的动物学家认为，鼠兔并非有害动物，也不是草地过牧的元凶；相反，鼠兔是一种有益动物，它们在维持健康的草地和生物多样性方面起着关键的作用。

由于体型小，鼠兔喜欢生活在植被非常低矮的草地。若草丛高而密的话，它们的密度相对很小。在过牧的草地上，它们最容易被看见，数量也最多，但是它们不是草地过牧的原因，而是过牧的后果。草地过牧的真正原因是人类过度放牧家畜，但是鼠兔很活跃，挖出的隧道像迷宫，所以很容易被当成草原退化的替罪羊。

实际上，因为鼠兔啃食好几种对牛有害或有毒的植物，它们有利于增进草原的质量。在没有鼠兔而且草又被过度啃食的区域，杂草生长茂盛，并开始替代草地。

鼠兔的另一个作用，是增加草地的通气性和通透性，使原本不多的雨水雪水都能渗入地下，滋养牧草，而不会从平坦板结地面上流失或在干燥的风中蒸发掉。

鼠兔的洞穴也为其他草原物种提供避难所或繁殖地。猫头鹰、地鸦、蛇和许多其他动物都利用这些洞穴。小沙狐甚至将鼠兔的洞穴扩大，变成自己的窝。

在中国，在不同的草地和高山地区所发现的鼠兔的种类不同。有些属于稀有或濒危物种。然而，不同种类的鼠兔都被贴上了有害动物的标签，并因此受到迫害。有必要号召终止这种无意义的屠杀，着手解决更困难的问题，即根据草地的生产量，决定家畜的数量，使中国广阔的草原实现最大可持续性畜牧生产。

和鼠兔一样，草原鼢鼠也是属于被消灭的对象。不同的是，鼢鼠是生活在地下的洞穴里，因此更难于捕杀。令人难以相信的是，鼢鼠的骨头可以代替虎骨入药，从而身价百倍。目前已经有人在人工饲养草原鼢鼠，来替代几近绝迹的虎骨资源。鼢鼠的药用价值可能有助于它的生存。对属于啮齿动物的鼢鼠来说，只要不是面对大规模的毒杀，它的生活习性、很强的繁殖力和对环境的适应，使它完全有能力应对其他形式的人为捕杀。

Not all pikas are grasslands species. The moupin pika is a small species that lives in sub-alpine forests up to high altitudes

并非所有鼠兔都居于草原。藏鼠兔就是生活在亚高山森林至高海拔的一种小动物。

Pheasants

With 52 species out of a world total of 196 China can boast to be the world centre of pheasant distribution. Pheasants are among the most spectacular of bird, being generally large with often long tails, brilliant coloration and extravagant ornaments such as colourful bare skin patches, ear tufts, brightly colored feathers, inflatable pouches, and iridescent patches on their plumage. The family is distributed all over China from the tops of mountains to the lowlands, from the tropics to the tundra.

Pheasants live on the ground and scrabble with their feet looking for seeds or insects or sometimes small plants to eat. Most nest on the ground but roost up in trees. Only the quails and some partridges huddle down on the open land to spend a cold night under the stars.

Tropical species include the green peacock whose male can fan out its tail and shake it in a stunning way during courtship revealing oscillated iridescent eye patches. The peacock species found in Xishuangbanna is the green peacock but in the wild it is virtually extinct however the peacock is a symbol of the culture of the Dai peoples of that region and in their efforts to reintroduce or domesticate peacocks from have introduced the blue Peacock from India which is an inappropriate species. It would be nice if people realized the distinction and domesticated or reintroduce the correct species.

The smaller but equally stunning gray peacock pheasant lives in the tropical hill forests. It also has iridescent eyespots on its plumage, but lacks the spectacularly large tail fan of the true peacock. The peacock pheasant emits loud double hoot calls that echo across the valleys.

The most widespread and perhaps globally best-known pheasant is the common pheasant which has 19 different races in different parts of China. This is the pheasant that has been introduced into Europe and North America for game shooting.

Some of China's pheasants are spectacularly beautiful. The Golden pheasant has a crimson chest, golden crown and the red mantle and bluish back wings. The make silver pheasant is a sleek elegant bird with white plumage boldly marked with black zigzag patterns. Even more spectacular is the gorgeous Lady Amhurst's pheasant with its long black-and-white barred tail set off by red basal feathers; and a blue back and green chest. The longest tail of all is found on the Reeve's pheasant. These tail feathers maybe up to one and a half meters long and are traditionally used in the flamboyance head dresses seen in Beijing Opera.

In the mountains we find the bronze-coloured monal pheasants, eared pheasants, snow cock and snow partridges all daring to live on barren hillsides up to over

Head of cock common pheasant is splendidly decorated. This is the premier game bird for sports shooting worldwide.

雄雉鸡的头部色泽精美。该鸟在世界各地都是被猎打的对象。

5000 m. Hardy little chukkor partridges live on the dry steepes and deserts of China and the most important pheasant of all is the tropical Reds jungle fowl. This is the ancestor of all domestic poultry.

Cock and hen Blood Pheasant rest on a rock before scurrying off to feed.

雄雌血雉共在一块岩石上小憩，正要起飞去觅食了。

Golden pheasants gather on the breeding grounds where males fight and display to attract the hens.

红腹锦鸡在繁殖季聚集一起，雄鸟争斗并展羽显示，以吸引雌鸡。

雉类

中国拥有全世界196种雉类中的52种，可称得上是世界雉类分布中心。雉类是最华丽的鸟类之一，多数有长尾巴，色彩鲜艳，外观上有华丽装饰，例如，耳羽簇、色彩鲜艳的裸皮，亮丽多彩的羽毛，可膨胀的袋状物，羽毛有具金属光泽的斑纹。雉类遍布全中国，从山顶到低地，从热带地区到冻原地带都有它们的身影。

雉类生活在地面上，用爪子在地上扒找种子、昆虫或小植物吃。多数雉类在地上筑巢，但在树上栖息。只有鹌鹑和一些石鸡在开阔的地面上偎集在一起，在星空下度过寒冷的夜晚。

热带雉类中有名的是绿孔雀。雄性绿孔雀在发情期间会将尾部开屏，绚丽夺目地摇晃着，展示着它们艳丽闪光的眼状斑纹。西双版纳的孔雀是绿孔雀，但野生绿孔雀实际上灭绝了。孔雀是当地傣族人的文化象征，人们想方设法引种孔雀或对它进行驯养，并从印度引入了蓝孔雀，但将蓝孔雀引到该地很不合适。要是人们能意识到两种孔雀间的区别或能驯养或引种适当的孔雀种类就好了。

灰孔雀雉体型更小，但同样美丽，生活在热带山林里。它的羽毛上也有色彩鲜艳的眼状斑纹，但是没有美丽的大尾巴，不能如同绿孔雀一样开屏。灰孔雀叫声洪亮，回音能响彻整个山谷。

环颈雉是分布最广同时也最为人所熟知的雉类，有19个不同的亚种，分布于中国的不同地区。这种雉类已被作为狩猎动物引入欧洲和北美洲。

有些雉类极其美丽。锦鸡有深红色的胸脯，金色的冠，背部与翅膀表面呈红色和翅膀背面略带蓝色。白鹇是一种纤细优雅的鸟类，拥有白色的羽毛，上面有鲜明的锯齿形黑色条纹。更引人注目的是华丽的白腹锦鸡，它有长长的黑白相间的尾巴，尾巴基部羽毛呈红色，背部为蓝色，胸脯为绿色。尾巴最长的雉类是白冠长尾雉，它的尾巴羽毛可长达1～1.5米，在京剧中通常用它作艳丽的头部装饰。

在山区有绿尾虹雉、马鸡、雪鸡和雪鹑，在海拔高达5000米的荒凉山区都能见到它们的踪迹。华丽的石鸡生活在干旱的峭壁和沙漠上。最重要的雉类是原鸡，它们是所有鸡的祖先。

Cock Blood Pheasant treads warily in search of food but shy of camera.

雄血雉轻轻踏步寻觅食物，但羞于相机，很难拍摄。

The male golden pheasant is the most boldly coloured bird in China. This may be the inspiration for the mythical phoenix.

红腹锦鸡雄鸟是中国色彩最艳丽的鸟。这可能就是神话中令人神往的凤凰。

The male of the silver pheasant has bold zigzag barring on its white upperparts but the female is a shy brown bird.

雄白鹇上体白色，具醒目的曲折横纹，雌性害羞，全身褐色。

The cock of Temminck's Tragopan is a colourful pheasant decorated with bold white spots. It feeds on the ground but roosts in low branches at night.

红腹角雉色彩浓艳，饰有大白点。它在地面取食，但夜间栖息于低矮树枝。

Male Reeve's Pheasant has black-edged golden scales and a boldly patterned head. It has tremendously long barred tail feathers.

白冠长尾雉雄鸟有带黑边的鳞状金片，头部有醒目图纹。尾羽有横纹，极长。

Cranes

Cranes are another family for which China is rightly famed. Nine of the world's 15 species can be found in China. These are splendid, tall, proud-looking birds and they are much appreciated in Chinese folklore for their longevity and the constancy of their relationships, with male and female cranes staying together for life. Some of the cranes are very rare. The Siberian cranes number less than 2000 in the world. They breed in northern Russia and fly south to China for the winter which they spend around Poyang Lake in the Yangtze Valley. The largest crane in China is the Sarus crane which occurs in Xishuangbanna in the extreme south of the country but this bird has become very rare now and has not been reported breeding in China for many years.

Another rare crane is the black-necked crane which lives on the high plateau, breeding in eastern Tibet and in the Rouergai marshes that straddle the borders between Sichuan, Gansu and Qinghai. The black-necked cranes fly over the great Himalayas to winter among wet valleys on the southern facies in Bhutan, NE India and NW Yunnan.

Perhaps the most famous, well known and well loved crane in China is the red-crowned crane which is almost as large as the Sarus. The red-crowned crane breeds in Jalong nature reserve and other sites in Russia and Heilongjiang but winters on the east coast of Jiangsu in the Yancheng marshes. This crane feeds in agricultural fields after the rice has been harvested. Here they can eat small plants, roots and fallen grain as well as many insects and aquatic invertebrates that live in the damp fields. A new change in agricultural practices is now threatening their winter feeding habits as local faremrs are switching to planting of cotton in stead of paddy rice. The cotton fields are not irrigated and do not leave much food for the cranes which have to wander wider and wider in search of good food.

The Siberian or White crane is one of the rarest birds in the world. The western population which used to winter in India is now extinct. Only about 2600 of the eastern population which winters around Poyang Lake, survive.

白鹤是世界上最稀有的鸟类。过去在印度越冬的西部鸟群现已灭绝。仅有约2600只在鄱阳湖附近越冬的东部鸟群尚存。

Close up of the red-crowned crane reveals a powerful bill that digs out plant and animal foods from wetlands and fields.

近看丹顶鹤就能看到它的嘴强有力，能从湿地和田野里挖掘出植物和动物为食。

The bold black and white markings of the red-crowned crane makes it a favourite subject among Chinese artists and photographers.

丹顶鹤醒目的黑与白色使它成为中国艺术家和摄影师都喜欢的题材。

鹤类

鹤类是另外一类中国引以为豪的鸟类。全世界15种鹤类当中有9种可在中国找到。鹤类是华丽、优雅和高贵的鸟类。在中国的民间传说中，鹤类是长寿的象征，也因雌雄鸟坚贞不渝的爱情而为世人所称赞。有些鹤类十分稀少，全世界的白鹤数量不到2000只。它们在俄罗斯北部繁殖，往南飞到中国的鄱阳湖过冬。中国体型最大的鹤是赤颈鹤，出现在西双版纳，但是这种鸟现在已非常稀少，已有多年未曾在国内报道过。

另一种罕见的鹤是黑颈鹤，它生活在高原地区，在西藏东部和跨越四川、甘肃和青海三省边境的若尔盖草地繁殖。黑颈鹤越过喜马拉雅山，飞到不丹南部、印度东北部和滇西北湿润的山谷中过冬。

在中国，丹顶鹤也许是最有名也最为人所熟知和喜爱的鸟。丹顶鹤几乎和赤颈鹤一样大。丹顶鹤在扎龙自然保护区以及俄罗斯和黑龙江的其他地方繁殖，但在江苏东部沿海的盐城沼泽地过冬。水稻收割后，丹顶鹤在农田里觅食，它们吃小植物、根茎、掉在田里的谷物，也捕食生长在潮湿田地的多种昆虫和水生无脊椎动物。如今，农民以棉花取代水稻的农业耕作方式正威胁着丹顶鹤在冬天的觅食习惯。棉田很少灌溉，丹顶鹤找不到足够的食物，只能被迫不断扩大自己的觅食范围。

Red-crowned crane on nest

就巢的丹顶鹤。

The black-necked crane only breeds in China.

黑颈鹤只在中国繁殖。

Wintering black-necked cranes enjoy the lake at Yunnan's Napahai nature reserve.

黑颈鹤在云南纳帕海自然保护区的湖泊上悠闲越冬。

Black-necked crane glides though the evening sunshine over Napahai Lake.

傍晚的阳光中，黑颈鹤从纳帕海湖上滑翔而过。

Waterfowl

China has no less than 45 species of waterfowl living on its lakes and waterways out of a world total of 147 species. These comprise three swans, 10 geese and a wide variety of ducks. Nine of these species breed north of China's territory and only visit as wintering populations. Some species are rare and endangered like the Lesser White-fronted Goose and Chinese Merganser. One species—the Crested Shelduck—has not been seen since 1971 and is now considered extinct.

Some of China's waterfowl such as the Ruddy Shelduck and Bar-headed goose are extremely hardy and can live on the salty lakes of the Tibetan Plateau. Others such as the dainty Cotton Pygmy-goose and Comb duck are tropical in distribution and only range into the extreme south of the country.

Huge flocks of some species can be seen in winter. Poyang Lake supports up to 50,000 Swan geese and 20,000 Greater White-fronted geese. 40,000 Tundra swans winter there also forming a spectacular scene like white petals floating. 100,000 Bean geese winter on the lakes of Anhui, together with huge rafts of mallard and other ducks.

One of the prettiest ducks and endemic to NE Asia is the mandarin duck. The male is resplendent with white eyebrows, a golden mane of long erectile hackles and cinnamon display 'sails' on its back. The mandarin nests in tree holes and may often be seen resting up in trees.

Two other local specialities are the Baikal and Falcated ducks both with shimmering green patterns on the head of the male, but the Falcated duck also shows off long plumed white-edged tertial feathers.

Ducks vary much in diet and habits. Some dive for their food, others dabble on the water surface and yet others come onshore to eat plant and animal materials. Bill shape is closely related to such habitats. The dabbling Shoveller has a huge spatulate bill whilst the fish-eating mergansers have a slender bill with a slight hook at the tip and serrated teeth along the bill length to help them catch and hang onto struggling fishes.

Even the ducks on the lakes and ponds of Beijing are well worth a look. Many are wild and their numbers are filled each autumn as migrating flocks make a stop over in the capital. Larger reservoirs around the city support larger flocks of ruddy shelduck, swans, pintail, mallard and associated herons and cranes.

Drake of pintail duck. Pintail winter in China in large numbers, mixing with other species in large flocks.

针尾鸭雄鸟。大数量的针尾鸭在中国越冬，它们与其他鸟混合成大群。

Drake Red-crested pochard in full breeding plumage.

赤嘴潜鸭雄鸟满身繁殖羽。

Male Mandarin duck checks out his good looks.

雄鸳鸯顾影自怜。

With every posture and movement, the gesture and Whooper swan exudes powerful elegance. It is larger than the similar coloured and commoner Tundra swan.

大天鹅的每一个姿态和动作, 都优雅非凡。大天鹅体型比小天鹅大, 但颜色类似。

水禽

中国的湖泊或水道里生活着全世界147种水禽中的至少45种，其中包括3种天鹅、10种雁和多种鸭。其中有9种水禽在中国北方境外繁殖，只在冬天在中国境内越冬。有些水禽是稀有物种，处于濒危状况，如小白额雁和中华秋沙鸭。另一物种冠麻鸭，自从1971年以来就没有看见过，现在被认为已经灭绝了。

中国的一些水禽如赤麻鸭和斑头雁的生命力特别强，能在青藏高原的咸水湖泊里生存。其他水禽，例如，秀丽的棉鸭和瘤鸭分布于热带地区，活动范围可以到中国的最南端。

在冬天，能看到大群的水禽。鄱阳湖养育着多达50,000只鸿雁和20,000只白额雁。40,000只小天鹅也在这里觅食过冬，如同白色的花瓣漂浮在水面上，构成了一道壮丽的景观。100,000只豆雁与大量的野鸭及其他鸭在安徽的湖泊越冬。

最漂亮的要数鸳鸯，它属于东北亚的地方种。雄性鸳鸯长相华丽，有白色的眉纹，齐刷刷直立的金色长羽毛，背部呈肉桂色。鸳鸯在树洞里筑巢，常在树上栖息。

另外两个当地特有种是花脸鸭和罗纹鸭，两种雄性鸭的头上都有发亮的绿毛，但是罗纹鸭也会展示长长的有白色边缘的拨风羽。

鸭的食物和习性多变。有的潜水觅食，有的在水边涉足，也有的在岸上吃动植物。鸟喙的形状与栖息地紧密相关。喜欢水底觅食的琵嘴鸭有一张扁平的大嘴，而食鱼的秋沙鸭有一张细长的喙，喙尖略呈钩形，喙的外缘长着细细的锯齿状，这有助于捕捉鱼并把挣扎的鱼牢牢夹住。

北京湖泊和池塘里的鸭也很值得一看，许多是野生鸭。随着成群迁徙的鸭在北京中途停留，每到秋天，鸭的数量都会大大增加。在首都周围的大水库能发现更大群的赤麻鸭、天鹅、针尾鸭、绿头鸭及相伴的鹭和鹤。

Drake mallard is resplendent in the winter months ready for courting as soon as spring is in the air.

绿头鸭在冬季华丽灿烂，准备春天一到便谈情示爱。

Drake of Falcated Duck.

雄罗纹鸭。

Ruddy Shelduck is a hardy species that breeds on the cold lakes of the Tibetan Plateau.

赤麻鸭耐寒，在西藏高原寒冷的湖泊上繁殖。

Butterflies and Moths (crypsis and mimicry)

Although not so fully studied and not so well known as the birds or mammals, the butterflies and moths of China are a joy and a wonder. They include some of the rarest and most beautiful examples in the world and with such great expanses still not affected by agricultural insecticides, this is one group of animals that can be said to still fare well in China and relatively easy to find and enjoy.

The butterflies of the temperate regions of northern China closely resemble and are often the same as widespread species that range across the whole of northern Europe and Russia. But China is enriched by an influx of tropical and subtropical species from the South and these reach surprisingly the far north during the warm summer months. Even in the north of Heilongjiang it is possible to find exotic tropical swallowtails and other large gaudy butterflies in the height of summer. Different mountain ranges support different local endemic species some of the more beautiful includes the *Apollo* butterflies. One of the rarest is the *Taeniopalpus aureus* butterfly found in Southeast China and a related species *T. imperialis* found in southwest time or the extraordinary swallowtail genus *Bhutanitis* which is one of the few protected insects in China.

It is in the tropical forests of the south that we find the largest and most showy examples. Giant silk moths can reach a width of 20cms. Whilst the gaudy black and yellow birdwing butterflies (*Troides*) can also attain an impressive size.

A walk in the forest can be a butterfly delight. Carpets of colourful species gather on damp ground, at the edge of puddles or especially on the salty ground where an animal has urinated. Some beautiful species even prefer to gather on the dung of elephants, monkeys and civets.

But why are some butterflies so keen to hide whilst others are so bold to show off their colours? Ask the birds and the lizards ! They could tell you that some butterflies are tasty whilst others are laced with bitter or even poisonous chemicals that make them most unappetizing. It is the distasteful species that dare to show off the brightest colours, the tastiest species that have to hide. Day-flying moths are almost all colourful and distasteful. But most moths are both plump and tasty, they wear camouflage and are inactive by day and only dare fly about in the relative safety of the dark. Now they must run the gauntlet of bats that can track them down in mid-air through their echo-location.

The Kallima butterfly looks like a leaf at rest but displays iridescent blue upper wings in its powerful flight when it is sure a bird cannot catch it. Less tasty Neptisbutterflies are boldly striped black and white and the Lacewing butterfly

A fritillary butterfly sips nectar from the flowers in a public garden.

Fritillary 蝴蝶在一个公园里吸吮花蜜。

Cehosia biblis is a colourful warning color of red, black and white.

But sometimes you can be fooled. You can catch a boldly coloured buttefly that you think you know and realize it is in fact a different species but coloured almost identically. This is what we call mimicry. Evolutionary forces have caused two species to look alike. Often one is a distasteful model wearing a bold pattern as a warning to birds and lizards that it is distasteful. The second species may be quite tasty but gets advantage by copying or mimicking the colours of the other.

Some butterflies get their toxic nature from the very plants they feed upon. The black and yellow birdwing butterflies feed on the leaves of a vine called *Aristolochia*. To most animals these leaves are poisonous. People even use this species for murder or to promote abortion. But the caterpillars of *Troides* are able to store the poisonous compounds without digesting them and store them in their own bodies to become poisonous in turn to their own potential predators. The poison is carried on through to the adult butterfly.

By late summer the forest skies are full of moths. The silk moth family boasts some large showy forms that gather around lights at night and provide a juicy breakfast for tits and jays when dawn breaks. They have feathery antennae that can detect a mate from several kilometers away. The fighter jets of the moth world are the hawk moths—Sphingidae. These are shaped for speed and fly long distances. They use river beds as motor highways to steer through the forest. The Death's head Hawk moth is distinguished by a skull-like visage on its thorax.

Yellow Eurema butterfly on yellow Aster.

Eurema 蝴蝶在紫菀花上。

Comma butterfly Polygonia c-aureum on summer daisies.

夏日雏菊上的黄钩蛱蝶。

Underside of the Red Lacewing butterfly is brightly patterned serves as a warning to birds that the species is distasteful.

红锯蛱蝶翅膀上醒目的色斑，对觊觎的鸟类天敌是一种警告。

The upper wing colouration of the lacewing butterfly Cethiosa biblis clearly fits it within the Common Milkweed mimicry complex.

草蜻蜓cethiosa biblis上翼的色彩，清楚地显示出它在色彩上模仿milkweed 类蝴蝶。

The common milkweed Danaus chrysippus form the basic model for a large complex of mimics.

金斑喙凤蝶是许多蛾类竞相模仿的主要对象之一。

The Kallima leaf butterfly at rest is marked just like a dead leaf but reveals orange and blue upper wings when it flies.

枯叶蝶休息时显如枯叶，但飞行时上翅膀显露出橙色和蓝色。

The upper wings of the leaf butterfly Kallima are more boldy coloured.

蝴蝶Kallima 翅膀上面的颜色更醒目。

蝴蝶和蛾类 （隐态和拟态）

　　尽管人们对蝴蝶和蛾类的了解与研究不如对鸟类或哺乳动物那样被研究透彻，但是蝴蝶和蛾类却受人喜爱，令人惊叹。中国有一些世界最罕见和最美丽的物种。它们分布广泛，但没有受到农药的影响。蝴蝶和蛾类在中国的生存状况依然良好，也易于找到和为人所喜爱。

　　中国北方温带地区的蝴蝶与整个欧洲北部和俄罗斯的蝴蝶相似或相同，这些蝴蝶活动于整个北方地区。但是，在中国的北方也有大量从南方来的热带和亚热带蝴蝶。令人吃惊的是，在炎热的夏季，它们出现在极靠北的地区，甚至在黑龙江的北部也能发现热带的凤蝶和其他艳丽的蝴蝶。不同的山脉养育着不同的地方性蝴蝶。最美丽的蝴蝶中包括阿波罗绢蝶。金斑喙凤蝶是最罕见的蝴蝶之一，见于中国东南部，与它相关的是西南地区的皇喙凤蝶和奇特的凤蝶，尾凤蝶属是中国少有的几种受保护昆虫之一。

　　最大、最绚丽的蛾类是大蚕蛾，见于南方热带森林。大蚕蛾翼展可达20厘米。另外，艳丽的黑黄相间的鸟凤蝶体型也很大。

　　在森林里行走可以是一个欣赏蝴蝶的过程。多种多样的蝴蝶聚集在潮湿的地面上，特别是在水坑边上或动物撒过尿的咸地上。一些美丽的蝴蝶甚至聚集在大象、猴子和灵猫的粪便上。

　　但是，为何有些蝴蝶喜欢躲藏，而另外一些蝴蝶却大胆地展示它们艳丽的色彩？问鸟类和蜥蜴吧。它们会告诉你有些蝴蝶是美味可口的，而另外一些却带有苦味或带有令它们倒胃口的有毒化学物质。味道不佳的蝴蝶敢于炫耀它们亮丽的色彩，而那些美味的蝴蝶必须躲藏起来。日间活动的蛾类几乎都是色彩斑斓且味道不佳。但是多数蛾类是肥胖和可口的，它们伪装自己，白天不活动，到天黑后相对安全时才敢出来。但是在晚上，它们必须逃避蝙蝠的攻击，因为蝙蝠能在空中利用回声定位捕捉到它们。

　　枯叶蝶在静止时看起来像一片树叶，但是在快速飞翔时，翅膀的上侧呈现出蓝色。口味差点的蛱蝶大胆地在身上长着醒目的黑白条纹。红锯蛱蝶具有红、黑、白等警戒色。

　　但是，人们有时可能被愚弄。当你捉到一只你自认为认识的显眼的红蝴蝶时，实际上可能是红色的另一种蝴蝶。这就是我们所说的拟态。进化的力量导致两个物种看起来相似。通常，艳丽的外表是口味不佳的蝴蝶用作对鸟类和蜥蜴的一种警示。其他可口的蝴蝶可能复制或模仿这些蝴蝶的警戒色。

　　有些蝴蝶能从它们取食的植物上汲取有毒物质。黑色的鸟凤蝶与黄色的鸟凤蝶以藤本植物马兜铃的叶为食。对多数动物而言，这些叶片是有毒的，人们甚至将这种植物用于谋杀和堕胎。但是，裳凤蝶毛虫能不消化它，而将这种有毒的物质贮存在体内，用作攻击潜在捕食者的武器。这种毒物能从毛虫一直传递到蝴蝶的体内。

　　晚夏时节，无数的蛾子在森林上空漫舞。蚕蛾体型很大，晚上聚集在灯光周围。在黎明时，通常会成为山雀和松鸦的美味早餐。它们有灵敏的羽状触角，在几公里外就能发觉配偶。天蛾是蛾类中的喷气式战斗机，它们的体型适合于快速远距离飞行。它们把河床当作穿越森林时的快速通道。骷髅头鹰蛾的显著特征是其胸前呈头骨状。

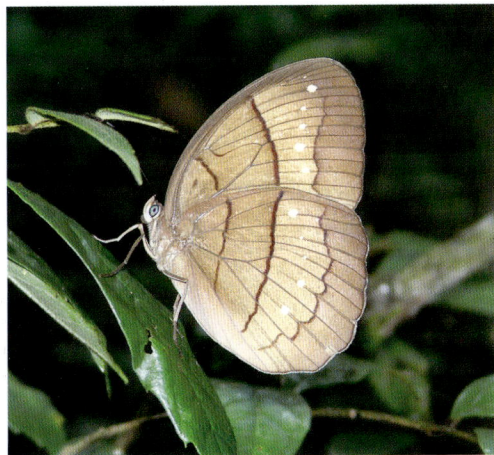

The fawn butterfly is a delicate creature of dark damp forests. It is one of a few Asian members of a large South American family.

纤细雅致的蝴蝶居于黑暗潮湿的森林。它是南美属蝴蝶的少数亚洲物种的一个。

A pretty Charaxes butterfly stops for a sip of fluid among the rocks of a dry riverbed. These are the fastest flying butterflies in the tropics.

一只美丽的螯蛱蝶在干河床上停下来，在岩石间吸吮液汁。这些蝴蝶是热带地区飞得最快的蝴蝶。

The long winged 'moon' moth (Actias) settles on a mango tree, displaying its long trailers. It is a relative of the silk moths and also makes a silk cocoon.

长翅膀的'月亮'蛾（Actias）在芒果树上安顿，显示出长长的尾部。它与丝蚕同属，也产蚕茧。

Butterflies huddle to sip rare minerals from a stream bed.

蝴蝶聚集在河床上，吮吸微量矿物质。

Dazzling hair-streak butterfly lives in southern limestone areas.

令人眼花缭乱的条纹蝴蝶生活在南部石灰岩地区。

The pretty small copper butterfly stands guard on an Aster flower, enjoying the evening sunshine.

漂亮的小铜蝴蝶站在紫菀花上，享受着傍晚的阳光。

Dragonflies

In China dragonflies and their larvae may be used in cuisine and as traditional medicine. But unlike in Western societies where dragonflies were usually regarded as malevolent creatures, in China they are always well respected. Dragonflies are regarded as an emblem of summer but also a sign of frailty and instability. In some areas they are regarded as spirits of the rice field and harbingers of good harvest. Dragonflies are a popular motif for making elaborate kites and may be found on many old paintings.

As with other taxa, China proves to be very rich in dragonfly species, though the group is not well studied and many more forms are likely to be yet described.

Many of the larger dragonflies are powerful flyers, they ride the typhoons and are distributed widely across the world. But the flimsier damsel flies are weaker flyers and tend to live along narrow streams, in forests or at the edge of ponds. Here we find many endemic species with narrow distributions and some rare species.

Dragonflies play an important role in freshwater ecology. The larvae are a dominant carnivore among the water weeds. They eat mosquito larvae and other pest species and the adults serve a similar controlling function in rice fields and urban areas. In addition the abundance and variety of dragonflies present is a good indicator of environmental quality. The numbers drop quickly a pollution levels rise. We can watch the dragonflies to know where it is safe to swim or catch fish. They are our friends and guides.

A mauve dragonfly watches from its hunting perch; dashing out to catch small passing insects that make up its crunchy diet.

一淡紫色蜻蜓从狩猎处观看着：猛冲出去捕捉过往的小昆虫，就有了它脆脆的餐食。

A libellulid dragonfly uses a lotus leaf shoot as a convenient hunting perch.

Libellulid 蜻蜓在嫩荷叶上狩猎食物。

Red pennant （Brachymesia furcata） dragonfly at rest on rocks.

红蜻蜓在岩石上小憩。

蜻蜓

　　蜻蜓大概是唯一一类对人类没有任何物质价值的动物，也从来没有被人饲养或人工繁殖过，这在中国是不可想象的，因为哪怕是臭名昭著的蛆都可能以"肉芽"的雅名进入菜单。也正是因为这样，我们对蜻蜓的喜爱是自然的、纯粹的、发自内心的。它生活在我们周围，也最亲近我们人类，它在大路上，在田园里自由穿梭与停靠。曾几何时，它也在乡间的屋前房后与农家的小顽童们捉迷藏，有谁没有童年在水边捕捉蜻蜓的美好记忆！它那硕大无朋的大眼睛，五彩缤纷的身体，在空中轻盈的舞姿，在水面上频频点水激起的串串涟漪，尖尖小荷上的倩影，让我们心旷神怡，让我们陶醉，从心底喜爱这自由的精灵。

　　在一些地区，它们被视为稻田的幽灵，预示着好收成。蜻蜓是制作风筝常用的主题，古画中也可见到。

　　中国的蜻蜓品种也非常丰富，目前有记载的约有350种，有人估计总数应该在600种左右。正如其他物种，但对蜻蜓的研究尚欠透彻，可能有不少种类尚未被描述。

　　许多体型较大的蜻蜓飞行能力很强，它们能随着台风飞，因而广泛分布在世界各地。但体型较纤弱的蜻蜓飞行能力较弱，往往沿着狭窄的溪流，在森林里或在池塘边生活。在这里，我们能看到许多分布较狭隘的特有种类和一些稀有种类。

　　蜻蜓对淡水生态有重要作用。生活在水草丛中的蜻蜓若虫是主要的食肉动物。它们吃孑孓和其他有害物种，成虫蜻蜓在稻田及城区起类似的服务控制作用。此外，蜻蜓的丰富度和种类也是一个环境质量很好的指数。污染程度上升，蜻蜓的数量就迅速下降。有蜻蜓在的地方我们就可以安全地游泳或捕鱼。它们是我们的朋友和向导。

Red pennant（Brachymesia furcata）dragonfly in flight.

红蜻蜓（Brachymesia furcata）在飞行。

Gilded damselfly graces the grass beside stream in Wuyishan Nature Reserve.

武夷山自然保护区，蜻蜓戏水溪边。

Laughing Thrushes

China can claim to be the global centre for a group of birds known as laughing thrushes. 37 species of the world total of 53 live in China. These are medium sized birds that live in groups, keep low in forest and scrub, have loud cackling group alarm calls which give the group its name and sharp piercing territorial calls. Several species have beautiful songs and it is for this reason that they are among the most treasured cage birds in China.

The Hwamei named for its beautiful white eyebrow is a common and very popular species, but the songs of the black-throated and spot-breasted are at least as melodious. Wildlife markets devote special space for the fancier of these song birds and many Chinese parks will have a bird corner where bird owners proudly hang their caged songsters in the early morning for some fresh air and to join the singing of other birds brought to the park.

Laughing thrushes occur up to high elevations with different species found in different habitats and regions across the country. Some species are boldly coloured with red or blue colours. Most are quite common but a few like the pretty yellow-throated laughingthrush are very rare.

The Peking Robin is a sweet-singing favourite cage bird in China.

红嘴相思鸟的鸣叫婉转悠扬，是一种常见的笼鸟。

噪鹛

中国被誉为噪鹛的起源中心。全世界共有53种噪鹛，其中37种见于中国。噪鹛体型中等，结群生活，栖于较低的森林和灌丛，受惊时发出大声咯咯叫声给鸟群报警－该鸟也正因得此名，占领地时发出刺耳尖叫。一些噪鹛的鸣叫婉转悠扬，成为在中国最受钟爱的笼鸟。

画眉因有美丽的白眉而得其名，该鸟常见，非常受欢迎，但黑喉噪鹛与斑胸噪鹛的叫声也一样悠扬动听。这些歌鸟在野生动物市场有专门的位置，中国许多公园也专门有一个角落，每天清晨许多养鸟的人都会聚在一起，把自己的鸟挂在树枝上，让鸟呼吸新鲜空气，来一场歌咏比赛。

噪鹛可以分布到高海拔的地区，在全国各地不同的生态环境中都可以发现很多种类。一些种类有红色或蓝色的醒目色彩。多数种类常见，但少数种类如黄喉噪鹛非常稀有。

White-browed Laughingthrush is a favourite songbird.

白眉噪鹛是最受人们喜爱的鸣鸟。

The female of the Fairy Bluebird is all cobalt blue. It has a melodious song that rings through the tropical forests of the southern provinces.

和平鸟雌鸟全身钴蓝。它歌喉优美，歌声响彻华南的热带森林。

Some of China's Special Plants

In addition to a long tradition of cultivating food plants and vegetables, China has also a long history of horticulture. This is the origin of tea, other camelias, roses, chrysanthemums and peonies. We owe much to the love of flowers shown by the imperial families of old and also to the monks in their temples and monasteries who cared for medicinal and ornamental plants of religious significance. We mention some of the special groups for which China is deservedly famous.

中国的一些特殊植物

Magnolia grandiflora has large elegant flowers when in full leaf.

广玉兰枝叶长满时，花卉大而优雅。

除了培育食用植物和蔬菜的传统外，中国还具有悠久的园艺历史。中国是茶的故乡，茶花、玫瑰、菊花和牡丹的起源地。我们应感谢宫廷对花的热爱，也应感谢寺院里的和尚，他们培育出许多有宗教意义的药用和观赏植物。我们接下来谈谈中国有名的特殊植物。

Magnolias

Magnolias are an ancient group of flowering trees. They evolved before bees existed and are designed to be pollinated by beetles. The flowers are simple but beautiful in their simplicity and creamy colours. They are distributed in many parts of the world but are most numerous in China where 160 of the world's 300 species occur. They are much admired in gardens but also used in traditional Chinese medicine.

The species *Magnolia officinalis* grows in the valleys of central China at middle altitudes. It has large leaves and a thick brown bark. It is this bark that has been valued for over two thousand years as a powerful medicine. Listed in the earliest medical texts as effective in treatment of hypertension as well as abdominal problems. The dried bark is till available in Chinese drug stores under the name 'houpu'. *Magnolia nitida* is used to make perfume.

Other species include the tropical form *Magnoila henryi* from the forests of Yunnan which can have enormous leaves up to metre long. At the other extreme the delicate *Magnolia stellata* is very small with starlike flowers.

Many magnolias come into bloom early in the spring before the leaves develop and thus make a splendid show of while to pinkish tulip-like flowers standing out clearly against the bare branches and dark forests.

木兰

木兰是开花乔木中的一个古老的类群。远在蜜蜂在地球上出现之前，它们的演化就已经完成，专门由甲虫完成授粉。木兰花形简单，但简单中透露出美丽，花色呈奶油色。木兰在世界上的许多地方都有分布，但是在中国的数量最多，占世界300种中的160种。它们常见于花园。木兰的干燥花蕾是一味中药，叫"辛夷"。

厚朴生长于中国中部中等海拔的山谷里。叶片很大，树皮为深棕色。两千多年来，其树皮一直被用作极具疗效的珍贵药物。据最早的医书记载，厚朴在治疗高血压和胃病方面具有很好的效果。在中药店里仍能找到该树皮，其名为"厚朴"。*Magnolia nitida* 用于制造香水。

其他木兰包括云南热带森林的大叶木兰，其叶子能长达1米。最小的木兰为星花木兰，其花为星状。

许多木兰在早春开花，先开花后长叶，花色从白色到粉红色，形状像郁金香，满树的花朵在光秃的树枝和幽暗的森林中显得格外的引人注目。

Delicate flowers of the magnolia brighten a dull wet day.

木兰花的雅致，照亮枯燥乏味的雨天。

Rhododendrons

Rhododendrons are a group of flowering bushes that live in the mountainous regions of Asia. They are prevalent through the Himalayas, most of the mountains of China and also in New Guinea. The greatest diversity of rhododendrons is found in SW China and in the Hengduan mountain range where more than 200 species can be found within a relatively small area.

Rhododendrons vary in size from the tree-like *Rhododendron arboruem* of the Himalayas to the tiny heath rhododendrons of the alpine zone and the delicate azalea bushes so loved by horticulturists. Many species have been collected and raised for gardens and some breeders have established showy hybrid varieties, but most rhododendrons are grown as wild stock and are showy enough in nature to deserve space in any garden.

They like acid peaty soil and are hardy evergreens. Many species grow as under-storey bushes beneath the sub-alpine conifer forests. Some have poisonous leaves so are disliked by China's cattle herdsmen. Other species have a poison in the nectar and honey and people can become ill or drugged by eating honey made by bees that visit these flowers.

At Longmenci in Dujiangyan, Sichuan there is a famous rhododendron garden where more than 100 varieties are kept. Another large collection is maintained in the alpine garden of Institute of Botany outside Kunming, Yunnan.

杜鹃

杜鹃是生长于亚洲山区的一类开花灌木。它们普遍见于整个喜马拉雅山脉（中国）的山区，以及新几内亚。中国的西南地区杜鹃资源最丰富，在横断山脉的一个不大的区域内有时竟能发现200多种杜鹃。

杜鹃的大小不一，有的高大如乔木，如长于喜马拉雅山脉的 *Rhododendron arboruem*，有的很小，比如高山地区的石南和园艺家们所喜欢的优美的杜鹃花。多种杜鹃被移植到花园里种植。有些育种工作者培育出绚丽的杂交品种，但是多数杜鹃是用野生种繁殖的，具有天然的美丽色彩，值得在花园种植。

杜鹃喜欢酸性泥炭土，属于耐寒常绿植物。许多

杜鹃生长在亚高山针叶林的灌木层。有些杜鹃的叶子有毒，因此中国的牧人不喜欢它。另有一些杜鹃的花露和花蜜有毒，人若吃了蜜蜂采自这些杜鹃花的蜂蜜会得病或中毒。

在四川省都江堰市龙门村有一个著名的杜鹃园，其中收集了100多个品种的杜鹃。另外一个大的杜鹃园是云南昆明附近一个植物研究所的高山公园。

A white rhododendron in early summer.

初夏的白杜鹃。

Rhododendrons are a favourite shrub plant in gardens around the world.

杜鹃花是世界各地花园都喜爱的一种灌木。

Pink rhododendrons in Dujiangyan.

都江堰的粉红杜鹃。

Buddha Bamboo is so-called because of its swollen culms.

佛陀竹因竹节膨胀而得名。

Bamboos

Bamboos are a family of giant grasses with strong hollow stems divided into segmented units called culms. They range from huge species in the moist tropics to fine 'arrow' bamboos of the sub-alpine forests on high mountains.

For centuries bamboos have played a significant role in Chinese culture. In the tropical regions bamboos provide a hundred uses in every day life. The form the poles for houses, the floors and walls and sleeping mats, fences, implements, chopsticks, musical instruments, baskets and even balls to kick.

Moso bamboo Phyllostachys pubescens is the most important commercial bamboo in China covering an area of 2.7 million hectares. The net value of its production was estimated in 1997 to be more than 1.3 billion $US, including a large export market. Moso is a large species and reputed to have the fastest growth rate of any woody plant—more than 1 metre per day. It provides a rich harvest of edible shoots but is also used for fencing, house building, furniture, medicine, many wood products and serves a great ecological surface of fast carbon sequestration and speedy restoration of degraded lands.

Bamboos are loved in Chinese gardens for their graceful form and the gentle soothing sound as their leaves rustle in a gentle breeze.

Some bamboos have the culms of their stems swollen into Buddha-like shapes.

The shoots of young bamboo make a wonderful vegetable or can be fermented to give special flavour to many dishes.Bamboos are the special food for the famous giant pandas.

Bamboos have a special place in Chinese painting either as a subject in its own right or as an element of landscape.

Bamboo is praised for the facts that it is hollow, meaning to be modest, and evergreen, meaning perseverance. So come the ancient catch word, I can live for three months without meat but cannot live for one day without bamboo in my garden.

竹子

竹子是大型草本植物，茎中空，有节。竹子的大小不一，大的如潮湿热带地区的巨龙竹，小的如亚高山林中的小箭竹。

几个世纪来，竹子在中国文化中发挥了重大的作用。在热带地区，竹子在日常生活的用途多达100多种，可用作房子的支柱、地板、墙、睡席、篱笆、器具、筷子、乐器、篮子，甚至可制成球用来踢。

在中国覆盖面积达 2.7 万公顷的毛竹是最重要的商用竹类。净价值的产量估计在 1997 年将超过 13 亿美元，其中包括一个大的出口市场。毛竹是一种巨大的、生长速度最快的木本植物－平均每天超过一米。它在提供了丰富的收获食用笋之外，也可用于击剑、房屋建筑、家具、医药、许多木材产品和服务一个伟大的生态表面的快速固碳和迅速恢复退化的土地。

竹子造型优美，微风吹过，发出轻柔而令人愉悦的声音，深受人们的喜爱，因而常常出现在中国庭院里。

有些竹子的茎上有节，膨胀成佛的形状。竹笋可做成美味的菜肴，也可腌制成许多其他菜肴的调味品。竹子还是大熊猫的专一食物。

竹子在中国的绘画艺术上具有特殊地位，要么是艺术的主题，要么是风景的组成部分。竹子中空，意味着谦虚，竹子常绿，意味着执着。有句古谚说：宁可三月不吃肉，但不能一日无竹。

A small bamboo factory makes chopsticks and combs for retail.

一个小竹器厂，生产竹筷子和梳子供零售。

Curtains of bamboo clothe the forest floor in SW China.

茂密的竹丛是中国西南森林里的林下植被。

Bamboo has a thousand uses.

竹有千用。

Sacred Plants

Many flowers in China have are revered as sacred and used in special ceremonies. In the autonomous prefecture of Xishuangbanna in SW Yunnan the Buddhist Dai minority preserve a suite of sacred plants in the gardens of their temples. These include *Bauhinia variegata* which is also the emblematic flower of Hong Kong Special Administrative Region. Another familiar tree is the pipa—Ficus religiosa with its long pointed apex. This the famous 'Bo' tree under which the Lord Buddha reached enlightenment. But this species was already revered by the peoples of South and SE Asia long before that, perhaps because when cut with a blade it bleeds a sap so reminiscent of human blood but pure white in colour. Another favourite is the sorrowless tree Saracca dives with its cheerful bunches of bright orange flowers. The rarest and most splendid of the sacred plants is the so-called 'golden lotus' *Musella lasiocarpa*. This is in fact a relative of the banana and lives in the forest, not in water. It has a large golden flower shaped of a lotus.

Water lotus is regarded as scared over almost all of China. Its clean simple shape inviting artistic efforts to capture its beauty in so many paintings and sculptures through the centuries.

A more northerly sacred plant is the ginko tree *Gingko biloba*. This is a very ancient tree and fossil ginko leaved, identical to those of today have been found with the dinosaurs of 200 million years ago. The species was thought by scientists to be long extinct until it was discovered in a Japanese monastery in 1691. More trees were found in Chinese monasteries and temple grounds, many very ancient.

The leaves are two lobed and some say they represent two entities that decided to become one, or the perfect balance between yin and yang. The leaves turn golden yellow in autumn, then fall quickly in the cool wind. Each leaf is said to utter a prayer as it falls the ground like golden snow. The fruits smell strange but the cooked seeds are a tasty food and have medicinal properties.

Sad-looking Paris lily with drooping cap. The species is valued as a medicinal plant.

下垂的花瓣使巴黎百合看起来凄凄惨惨。但该物种可是宝贵的药用植物。

神圣的植物

中国的许多花卉被视为是神圣的，并被用于特别的仪式。在滇西南的西双版纳自治州，傣族的佛教徒在寺庙的院子里种植很多的神树，其中包括紫荆花（宫粉羊蹄甲），它是香港特别行政区的区花；还有为人们熟知的，具有长尖花序的菩提树，佛祖就是在菩提树下修成正果。但是在佛教出现之前，这种树就已经为南亚和东南亚的人们所敬奉，因为用利刃砍在树上，它会像人体流血一样流出白色的树液。另外，还有人们喜爱的无忧树，其花多而密，红似火焰；其中最罕见，最华丽的神树是地涌金莲，它实际是芭蕉的亲缘物种，生长在森林里，而非水中。地涌金莲花形巨大，花色金黄。在整个中国，荷花几乎都被认为圣洁的代表。荷花清洁无瑕，它的美丽被反映在几个世纪以来的许多油画和雕塑作品中，成为佛教的象征。

银杏是生长在更靠北地区的神树。它是一种古老的树种，与恐龙一样出现于2亿年前，化石银杏的叶化石与今天我们所见的银杏叶相同。在1691年日本僧侣发现银杏前，该物种曾一度被科学家们认为早已灭绝。后来，在中国的修道院和寺院里又发现许多古老的银杏。

银杏叶有两个裂片，有人说这代表两个个体合二为一，或代表阴与阳的平衡。在秋天，叶片变黄，随寒风飘落。据说，每片叶子在飘落时都会许一个愿。银杏果实的气味奇怪，但炒熟的种子是美味的食品，也可作药用。

The sacred golden lotus banana Musella lasiocarpa of southwest Yunnan is regarded as magical because of its resemblance to the flower of the lotus and also its medicinal food properties.

滇西南的地涌金莲被视为有神效，因为它外形似荷花，可药用。

Medicinal plants

China has one of the oldest and most comprehensive traditional medicine knowledge in the world. Ancient books dating back almost three thousand years list and describe the many herbs and even animal parts that form the basis for traditional remedies. Several thousand plant species are listed, including 1,500 in Yunnan province alone.

Favourite medicinal plants include fritillary bulbs and gentians collected in the high mountains, the roots of ground orchids *Gastroidea elata* and some bracket fungi collected in the forests, the bark of some trees, many dried fruits, berries and leaves. One of the strangest and most expensive medicinal plants is called 'summer grass - winter worm' *Cordyceps agrestis* and is in fact not a plant at all. It is formed when a fungus attacks and takes over the larvae of a moth that burrows into the alpine soils of the Tibetan Plateau. These mummified 'worms' are regarded as a traditional cure-all that helps fight everything from AIDS to cancer and ageing. Collecting the 'worms' has become a lucrative industry for many villagers in the summer months.

Although many plants are recommended for specific ailments—worms, fever, abdominal pains, headaches, broken limbs, wounds, childbirth, sexual dysfunction and the like, the application of TCM is not so direct as Western medicine. Chinese medicine is a more holistic approach. Drinking infusions of the correct mixture of herbs is believed to restore overall health by regulating the balance between hot and cold—the yin and yang and the level of the 'chi' or life spirit.

Thornapple or moonflower (Datura) is a somewhat poisonous member of the potato family. The black seeds are encased in a thorny green pod and the flowers are long and trumpet-shaped. The name moonflower came from the habit of the flower opening periodically through the night to release a powerful sweet scent but the flower is grown through China and many other countries because of its use as a halucogen and a narcotic. All parts of the plant are poisonous but taken as a tea or in small quantities can induce a pleasant drunkenness and strangedreams. The plant has been used for centuries in China as an anaesthetic to dull the pain of surgery.

Many other valuable medicinal plants are gathered by villagers from the forests and mountains, dried in simple ovens or under the sun and then sold through traders, or in markets for home use or to factories of TCM products or prescribers of such medicines.

Price is determined by supply and demand but there is little control over-collecting activities and in some areas over-collection has led to rarity or even local extinction. Efforts are now being made to reduce the pressure on endangered species by a combination of approaches. These include establishing protected

The sacred lotus has been revered by many cultures and religions.

莲花在许多文化和宗教中都被视为神圣的植物。

areas, limiting collection to certain seasons, collecting areas or quotas, reducing wastage in the collection and drying stages, identifying substitutes for rare species and rearing of some precious species in home gardens rather than gathering entirely from the wild.

One valuable medicine is Ammomum villosum, a wild ginger of the tropical forests in southern Yunnan. The seed pods are harvested and dried for use in a variety of medicinal recipes. The plant requires moisture and shade. It does not fare well in open home gardens so is planted and encouraged under wild forest trees, even well inside nature reserves. This primitive form of farming involves merely weeding around wild plants to give them more space to reduce competition from other wild plants as well as a bit of planting out of split plants into new sites. The villagers harvest the seed pods in due season and make a nice addition to their generally otherwise tiny income.

Ginseng is a forest herb of NE China. The most sought species in *Panax ginseng*. The root of ginseng is strangely deformed into fantastic shapes but often looking like a bearded man. It is this 'man root' that is used for medicine, particularly for improving male sexual performance. Ginseng has been cultivated in Korea and China for at least 2000 years, though most practitioners believe the wild plants are more potent and thus more valuable. Cultivation is difficult because of the slow growth, prevalence of diseases and need of shade and humidity. Modern ginseng farms are made inside the forest under plastic sheets. A new technique of growing ginseng in hydroponic tanks allows for much faster growth and apparently higher levels of effective compounds in the resultant plants.

Some wild fruits are regarded as medicinal and spiritual. The orchard is associated with the idea of an aloof and righteous gentleman.

The residence of a famous doctor is called apricot orchard. Legend goes that in the past a very famous doctor practiced medicine far and wide. For each patient healed, his demand was to plant one or more apricot trees by his house. Finally his house was buried among a dense forest of apricot trees.

Legend goes again that Confucius liked to give lectures to his disciples under apricot trees, thus the term Apricot Forum is used to denote outdoor teaching.

药用植物

中国有世界上最古老、最综合的传统医学知识。中国的医学古书几乎可追溯到3000年前，其中罗列和描述了许多草本植物和动物，它们构成了传统医学的基础。罗列的药用植物多达几千种，其中仅云南省就有1,500种。

人们喜爱的药用植物包括生长在高山地区的贝母和龙胆，天麻和灵芝，一些树皮，许多干果、浆果和叶子。冬虫夏草是最奇怪、最昂贵的药用植物之一，它实际上根本就不是植物。它是真菌寄生于一种蛾子的幼虫后形成的，这种蛾子的幼虫穴居于青藏高原的高山土壤中，这些干瘪的"虫子"被认为是传统的百宝丹，有助于抵抗包括艾滋病、癌症和衰老在内的各种病。夏天，采集冬虫夏草已经成为一个利润丰厚的产业。由于过度采集，冬虫夏草的野外资源已经枯竭。

尽管许多中药只能治疗某些特定的疾病，如寄生虫、发烧、胃疼、头痛、骨折、外伤、分娩、性功能障碍及类似症状，但中药仍然不如西药的针对性那样强，但更多的是一种整体机能的调理。中医认为，喝正确配方的中药能调节热和寒（阴和阳）之间的平衡，有利于调气。

蔓陀罗或月光花是马铃薯属的一个有毒品种，其黑色种子包藏在带刺的绿色荚内，花很长，呈喇叭状。月光花夜晚按时开放，释放出强烈的甜香味，该花因此得名。因其迷幻和麻醉作用，月光花在中国及一些其他国家都有栽种。月光花全株有毒，少量饮食可引发类似微醉或作梦的愉快感。中国几个世纪以来都使用该植物作麻醉剂，以减轻手术时的疼痛。

许多宝贵的药用植物是由村民从森林和山上采集，在简单的炉子上烘干或在太阳底下晒干，然后出售给药商或中药厂。

中药的价格由供需决定，但是采集几乎不受管制。在有些地区，由于过度采集，一些物种已经很稀少，甚至在局部地区已经灭绝。现在正在综合采取多种措施减小濒危物种灭绝的压力。其中包括建立保护区、季节性限采、划定采集区、限定采集量、减少采集和干燥阶段的损耗、寻找稀有物种的替代物，以及人工种植一些珍贵物种，而非完全从野外采集。

砂仁是云南南部热带森林中的一种野姜，是一种宝贵的药物。它的荚果采集晒干后，是一种用途广泛的中药。这种植物生长在潮湿、阴暗之处，在空旷的庭院里长不好，应种在野外森林里，在自然保护区里也能长好。只需要采取原始的耕作方式，将它周围的杂草除去，增加其生长空间，减少其它植物的竞争，为了利于其生长，可把一些植株分开，种到其它地方。村民在适当的季节采集荚果，能使他们原本微薄的收入得到可观的增长。

人参是中国东北地区的一种林下草本植物，也是一种需求最大的中药材。人参造型奇特，看起来象一个留有胡须的人。正是这人形的根具有药用价值，它尤其对提高男性性功能具有效果。在朝鲜和中国，人参的种植历史已经至少有2000年，大多数医生认为野生人参有效，因而也更昂贵。人参的培育很困难，因为它生长缓慢，易染病，且需要阴暗和潮湿的生长环境。现代人参种植场都建在森林里的塑料篷下。利用培养液能使人参长得更快，且明显含有更多的有效成分。

某些野果被认为有医学或精神作用。兰花象征着君子的清高与正直。

有一著名的大夫居址就叫做杏花园。传说古时候有位著名的大夫四处行医。每逢一位病人治愈，他就叫这位病人在他家院子周围种上一棵杏树。最后，他家就深深埋在杏树林中了。

传说孔夫子也爱在杏树下给他的子弟们传经说道，因此"杏坛"这一词就用来描述课外学习。

Fruits of the sacred Gingko are used as food and medicine.

银杏的果实可以食用，也能入药。

Antiquity of Chinese Traditional Medicine

"Shen Nong's Herbal Classic", or "Herbal Classic" for short, is one of the earliest medical monographs, and was compiled in the year of 200 AD the last year of Eastern Han dynasty.

The Classic lists 365 kinds of drugs, 252 kinds of herbs, 67 types of animals and46 sorts of mineral drugs. It details every drug, elaborating its smell and attributes, its function, other names and living environment. Based on the performance and effects of the drugs, they are divided into 3 classes, upper class, middle class and lower class. The upper class of 120 drugs are nontoxic, most of them being nourishing, such as ginseng, licorice, Chinese fox-glove root and Chinese dates, etc., which can be used in daily life. The Middle class also consists of 120 types, some toxic and some not; some of them nourish the primary qi greatly and relieve unease of the body, such as lily, angelica, longan, and pilose antler, etc; some eliminate the pathogenic factors and improve immunity, Today, Chongqing and other regions in China have begun large-scale artificial cultivation of Artemisia annua, used to extract artemisinin.

Huatuo, a medical scientist at the end of the Eastern Han Dynasty, from Qiao County of the Pei Kingdom (now Bo County of Anhui Province) invented anaestetics.

Narcotics were used before the birth of Huatuo, but were only used in warfare, murder, etc, but not for surgery. Huotuo summed up the former experiences of narcotics and noticed that people went into death-like sleep when drunk, so invented the narcotic method of taking anesthesia with alcohol. This method greatly enhanced and expanded surgical practice. Huatuo would ask patients, who failed with acupuncture and herb tea, to take anesthesia, and start surgery when they became unconscious.

More than 1,500 years ago, Chinese doctors began to use Artemisia apiacea to treat malaria. In fact, Artemisia apiacea does not contain artemisinin, but Artemisia annua and Artemisia apiacea are in the same genus, most likely, the Artemisia apiacea collected were mixed with Artemisia annua which is effective against malaria.

Today, Chongqing and other regions in China have begun large-scale artificial cultivation of Artemisia annua, used to extract artemisinin.

Variety of products has been the cultural backbone of farming and eating in most parts of China. It is also an ecologically and economically sound approach. The modern trend of going for high yields of a single crop is a gamble and its sustainability untested.

很多产品已成为中国大部分地区劳作与饮食文化的支柱，也是生态上和经济上很健全的做法。单纯追求某一作物的高收成这一现代做法，是对自然的赌博，其可持续性还未经过考验。

传统中药的发展

《神农本草经》，简称《本经》，是我国现存最早的一部药学专著。一般都认它是东汉末年（约公元200年）之作品。

《本经》载药365种，其中有植物药252种，动物药67种，矿物药46种。《本经》对每味药所记载的内容，有性味、主治、异名及生长环境。根据药物的性能和使用目的，分为上、中、下三品。上品一百二十种，无毒。大多属于滋补强壮之品，如人参、甘草、地黄、大枣等，可以久服。中品一百二十种，无毒或有毒，其中有的能补虚扶弱，如百合、当归、龙眼、鹿茸等；有的能祛邪抗病，如黄连、麻黄、白芷、黄芩等。下品一百二十五种，有毒者多，能祛邪破积，如大黄、乌头、甘遂、巴豆等，不可久服。

《本经》的问世，对我国药学的发展影响很大。历史上后来编著的许多具有代表性的《本草》都是源于《本经》而发展起来的。一些治疗方法，例如麻黄平喘、黄连治痢、牛膝坠胎、海藻治瘿瘤等，不但确实有效，而且有一些还是世界上最早的记载。如用水银治皮肤疾病，要比阿拉伯和印度早500～800年。

华佗（约145-208）东汉末医学家，沛国谯（今安徽亳州市谯城区）人，发明了麻沸散。

利用某些具有麻醉性能的药品作为麻醉剂，在华佗之前就有人使用。不过，他们或者用于战争，或者用于暗杀等，真正用于动手术治病的却没有。华佗总结了这方面的经验，又观察了人醉酒时的沉睡状态，发明了酒服麻沸散的麻醉术，从而大大提高了外科手术的技术和疗效，并扩大了手术治疗的范围。华佗治病碰到那些用针灸、汤药不能治愈的腹疾病，就叫病人先用酒冲服麻沸散，等到病人麻醉后没有什么知觉了，就施以外科手术。

1500多年前，中国人就开始利用青蒿（Artemisia apiacea）来治疗疟疾。其实，青蒿中并不含有青蒿素，由于青蒿与黄花蒿是同属的植物，很可能在所谓的青蒿中混有黄花蒿，从而具有一定的疗效。

目前在中国重庆等地区已经开始大规模人工种植黄花蒿，用来提取青蒿素。

Gentians form gem-like stars on the forest floor but the plants are harvested as traditional medicine.

龙胆在森林地里星星点点如宝石，该植物常被采来用作传统药材。

Red precious. The ripe seeds of the ginseng plant. Each can be raised into a valuable new plant for traditional medicine.

红色之宝·人参植物的成熟种子。每一粒种子又可培育出一棵宝贵的新人参，做成为传统药物。

The scaly ant-eater or pangolin is a creature of the tropical forests of the south. It climbs up trees at night to open and raid ants nests and termite nests, licking up the grubs and adults with its long sticky tongue. Sadly many people believe the scales have medicinal properties and this animal is being heavily hunted and killed for the medicine trade.

身带鳞片的穿山甲是南部热带森林地区的一种动物。它在夜间爬到树木开口处劫掠蚂蚁巢和白蚁巢，用其长而黏的舌头舔食蛴螬及蚁类。不幸的是许多人认为，穿山甲的鳞片有药用性能，因而该动物被大量捕杀。

A herbalist in a Dai minority village practices traditional medicine using ancient texts and herbs he collects himself in the surrounding hills.

一位傣族民间医生，根据古代文献，用自己在周围山上收集的药材，从事传统医药活动。

PART FOUR
Threats to China's Biodiversity

Challenges

For centuries we have enjoyed the benefits of nature's bounty, accepting it as a free gift with little consideration to its sustainability and little concern that the supply may dwindle or collapse.

Our ancestors have cut into the great forests to get the timbers needed for housing or fuel. They have replaced the huge herds of natural ungulates on the extensive grasslands with ever growing herds of cattle, horses, sheep and goats.

Hunters and fishermen have taken their harvest of wild creatures to eat or sell. We have worn the furs of large mammals, dined on the meats of birds and fishes and scoured the land for tasty fungi, bees honey, fruits and vegetables. The most useful species have become domesticated for everyday use around our farms and villages.

China is the origin of the yak, cattle, horses, pigs, chickens, sheep, and ducks. Even the insects have been domesticated for centuries to give us the modern honey bee and silk worm. China's nature has offered us wild rice, millet, mangos, citrus fruits, apples, pears, persimmons, litchis and kiwi fruit.

Centuries of China's doctors have learned and recorded the properties of many plants upon our bodies so that China today has the most extensive and complete traditional medicine system in the world. More than 15,000 species have been listed as having useful medicinal properties and by combining traditional uses with modern scientific testing and synthesis, we get maximum benefit of both the old and new.

But China's great success has brought its own threats. Our population grew and grew so that China has become the most populous nation on earth and contains almost one quarter of all the people on the planet. The natural resource base that we have taken so much for granted is suddenly reaching its limits. There are few remaining forests to cut and they cannot meet today's enormous demand. China is now a major importer of timber from other countries. There is no new agricultural land to clear and we have to rely on science to push up the unit production per hectare of existing farmland. There is not enough game or fish to grace so many tables and not enough wild plants to meet all our medicinal needs.

Not only has such insatiable demands caused many of our native species to become rare or endangered but China has now become a major importer of timber, wildlife, fish, furs and other natural products and is depleting the natural stocks of neighbouring and even far distant countries.

We can say that over-harvesting of species and clearance and fragmentation of natural habitats are ancient but growing threats on the natural environment. But new threats have now arisen that add to the pressures on our beleaguered natural treasures.

第四章
中国生物多样性所面临的威胁

挑战

几个世纪以来，人们一直都在享受着大自然所带来的恩惠，把它当作免费的礼物，而不考虑其可持续性，也不关心它的供应有朝一日会减少或崩溃。

我们的祖先走进大森林伐木建房，或采集薪柴，并用越来越多的牛、马和羊取代广袤草原上的大群的天然有蹄类动物。

猎人和渔夫捕捉野生动物食用或出售。我们穿哺乳动物的毛皮，食用鸟类和鱼类的肉，在土地上寻找美味的真菌、蜂蜜、水果和蔬菜。将最有用的物种在农场或村庄附近驯养，以供日常之需。

中国是牦牛、牛、马、猪、鸡、羊和鸭的起源地。甚至有的昆虫也被驯养了几个世纪，如现代的蜜蜂和蚕。大自然还为我们提供了野稻、稷、芒果、橘子、苹果、梨、柿子、荔枝和猕猴桃。

几个世纪以来，中国的医生学习和记录了许多植物对身体的影响，从而，在今天，中国有世界上最广泛、最完整的传统医学，用作药物的物种达15,000种以上。通过将药物的传统用途与现代测试和合成技术相结合，相得益彰，使我们能获得最大的效益。

但是，中国在取得巨大成就的同时也为自己带来了威胁。我们的人口在增长，中国已经是世界上人口最多的国家，约占世界人口的四分之一。我们曾经视为用之不竭的自然资源基础突然到达了极限。剩余的森林已很少，已不能满足今天的巨大需求。中国如今是一个木材进口大国。没有了可供开垦的土地，必须通过科技进步来提高现有耕地的单位产量。已经没有足够的猎物或鱼来满足众多的餐桌，也没有足够的野生植物来满足对药物的需求。

无止境的需求不仅导致许多乡土物种越来越少或者濒危，也使中国如今成为木材、野生动物、鱼、皮毛和其他天然产品的进口大国。可以说，对物种的过度采集和天然栖息地的清除和破坏是由来已久，但对自然环境带来的威胁在不断加大。新的威胁进一步加大了对自然财富的压力。

Pollution, Acid rain, Damming, Communications

China's fast growth of industry has not always been accompanied by highest environmental standards. Mines tip their spill into waterways, factories discharge their waste and millions of tall chimney stacks belch black clouds into the air from the coal-fired furnaces.

Growing use of agricultural pesticides and fertilizers leads to chemical run off into waterways both poisoning biota and creating dangerous growth of green algae as a result of eutrophication.

Discharge of SO_2 into the atmosphere from factory discharge and car exhausts leads to acid rain which may kill fish in waterways and trees in forests. In the steep valleys of southern China, acid clouds hang about for days before being washed down in rain. Chongqing and Ganzhou are listed among the towns most affected by acid rain in the world.

Pollution is a hazard to human health, food production and wildlife alike. It has been the price to pay for fast economic growth. China is not unique. This was exactly the path already followed by most developed nations in Europe, America and Japan. Develop first, clean up later. The pollution threat in China is now starting to get serious attention by the government and strict laws on pollution, emissions, discharge and environmental impact assessments are coming into place to improve the situation.

China has massive potential for development of hydropower and also has great needs of water for its irrigated agriculture. This has led to the construction of literally thousands of dams and weirs across the country. Some big projects like the Three Gorges and Gezhouba dams on the Yangtze receive a lot of publicity and comment, but most dams are quietly built by local agencies in the smaller valleys of the great rivers with little attention to their environmental impacts.

Dams bring both good and bad effects. They bring economic benefits in terms of clean energy production, water supply for irrigation and drinking through the dry season, and a check on floods. Their efficiency and investment put a greater emphasis on good protection of the upstream water catchment on which they in turn depend for their water source. This sometimes leads to reforestation, eco-logical restoration and better forest protection. But on the negative side, such

Cows help recycle rubbish outside Lhasa.

拉萨郊区的垃圾牛。

dams block critical pathways by which aquatic species move from upstream spawning areas to downstream feeding grounds. The reduced wet season water flow means many downstream wetlands are no longer replenished with new water and silt as they were before. This causes loss of fisheries, feeding grounds for wintering waterfowl and other wetland fauna and flora.

Just as dams fragment and degrade the aquatic habitats of China, so the constriction of roads, railways, pipelines and the agricultural encroachment that accompanies new communications lines, cut across and fragment other natural habitats such as forest, grasslands and major wetlands.

As habitat patches become smaller and more isolated so they start to lose many of the larger species that play keystone roles in their ecological stability and equilibrium. Carnivores vanish when there is not a large enough prey population to live off. Ungulates vanish as they are too vulnerable to human disturbance and hunting in small patches, larger birds and ground living birds soon follow together with primates and larger squirrels. China has many 'dead' forests where at first sight the habitat looks quite good but when one looks in detail, the wildlife is mostly absent. Without ungulates to eat them the pattern of recruitment of young trees is quite changed and without fruit eating primates and birds the distribution of seeds is also warped. Gradually the tree composition changes and always in the direction of reduced species richness and reduced ecological service efficiency.

污染、酸雨、水坝和交通设施建设

中国工业的快速增长并不总伴随着更高的环境标准。矿山直接将尾渣弃置在河道里，工厂排放废物，上百万个高耸的烟囱将煤炉子里放出的黑烟排入大气中。

农药和化肥的使用量越来越多，并随雨水流入水道，不仅毒害了生物群，而且富营养化的出现促进了绿藻的生长。

来自工厂废气和汽车尾气的二氧化硫形成了酸雨，酸雨会杀死河水中的鱼和林中的树木。在中国南部的陡峭的峡谷中，酸云会在空中滞留好几天，才会被雨水带到地面。重庆和江西赣州被认为是世界上酸雨危害最严重的城市。

污染会危害到人类的健康、食物的生产和野生动物的安全。快速的经济增长需付出代价，中国也不例外。这也是欧洲、美国和日本等发达国家走过的那条先开发、后治理的道路。中国的污染威胁如今开始受到政府的高度关注，为了改善环境，已经制定了关于污染、废物排放和环境影响评估方面的严格法律。

中国的水电资源极其丰富，也需要大量的农业灌溉用水，因而在全国建立了大量的水坝和堰。一些大工程，诸如三峡工程和葛洲坝电站等，受到了大量的宣传和评论。但是，那些地方机构在小山谷中建立的众多水坝的环境影响却鲜为人知。水坝既有好的效应，也有坏的效应。它们能带来经济利益，包括清洁能源生产，在干旱季节供应饮用和灌溉用水，以及阻止水灾等。为了确保可靠的水源供应，水电会促进上游流域的保护，包括造林、恢复生态和更好地保护森林。从消极面看，水坝阻断了水生物种在上游的产卵区与在下游的觅食区间的关键洄游通道。在雨季的径流量会下降，意味着下游湿地的水源和淤泥补充也将减少，这造成了渔业的损失及越冬水禽和其他湿地动物采食地的丧失。

如水坝使中国的水生栖息地斑块化和退化一样，建设公路和铁路，铺设管道和农业的蚕食也会将诸如森林、草地和大湿地等天然栖息地切断、斑块化和建立起新的通道线路。

当栖息地斑块变得越来越小，越来越孤立，将失去许多在维持生态的稳定性和均衡方面有重要作用的大型物种。当没有够多、够大的猎物种群时，食肉动物就消失了。当斑块脆弱到不能承受人类活动和打猎的干扰时，有的蹄类动物就会消失，大型鸟类和在地面活动的鸟类连同灵长类和更大的松鼠将相继迅速消失。中国有许多"死"森林，第一眼看上去，栖息地看起来相当好，若仔细瞧瞧，会发现几乎没有野生动物。没有有蹄类动物的啃食，树木更新的方式就改变了，没有吃水果的灵长类和鸟类，种子的传播也受到限制。逐渐地，树木的组成改变了，且总是朝着减少物种丰富度和降低生态服务效率的方向改变。

Pollution is a huge problem in China. The discharge of several million factories flows into waterways.

污染在中国是一巨大问题。数百万工厂的废水排入水道。

Thousands of dams in China pose hazards to wildlife and fragment freshwater habitats preventing free passage of fish and amphibians.

中国数以千计的堤坝不仅对野生物构成了威胁，而且阻断了淡水生境，使鱼类和两栖动物无法自由通行。

Alien Invasive Species

Many non-native species have been deliberately introduced into China, often with great positive results. These include foods such as potatos, chilies, grapes, tomatos, corn; also commercial crops such as palm oil, rubber, coffee, timber trees and ornamental plants to adorn our gardens or fish to raise for the market. In most cases such introductions are harmless and do not threaten local biota, but also quite commonly such introductions can cause great damage.

Whereas we used to recognize habitat destruction and over-harvesting of wild species as the greatest threats to wild species populations we now recognize a new and more insidious threat creeping across the globe and especially in those regions where human activity are causing fast and dynamic changes to natural ecosystems. This is the threat of alien invasive species.

Simply stated an alien invasive species is a species that has got a foothold in a habitat where it was not originally present but where by virtue of its greater vigour, faster reproduction or greater competitive edge the new species spreads aggressively causing damage to the newly colonized habitat or out-competing and displacing native fauna or flora.

An exotic or alien species often has a competitive edge on first colonization in a new habitat simply because there are no diseases or herbivores present in the new ecosystem that are adapted to eating or controlling that species. It its own original habitat each species is kept in balance by diseases or predators that become more intense with high density. Removed from these natural constraints the species can spread as a pest until such time as its own predators can also colonize the new ecosystem or other species or pathogens within the new eco-system adapt to exploit this new food source in their midst.

The spread of potentially invasive species has been dramatically increased through the logarithmic increase in human trade and movement. Organisms and seeds get carried across the world in ships and planes, dumped in ballast water, held in frozen foods, living in wooden crates, seeds mixed in with grain shipments or imported grass seed, carried on the clothes and shoes of international travelers or hidden in soil samples attached to living plant material. Even more disastrous has been well meaning but catastrophic deliberate introductions of plants or animals for horticulture, agricultural trials , food or misguided efforts at natural pest control.

China is particularly vulnerable to invasion by such alien species because its wide range of available habitat conditions probably offers some suitable home to almost any species that finds its way into the country. The fast pace of change to the landscape leaves huge areas of open habitat inviting rapid colonization and China has rapidly become the single largest importer and exporter of trade goods

Although pretty and often planted in gardens, Lantana camara is a fierce weed that spreads across pastures, is difficult to remove and not eaten by cattle.

马缨丹虽然漂亮, 而且常常被种植在花园, 但它却是一种入侵性强的杂草, 蔓延全国各地的牧场, 很难去除, 而家禽也不吃这植物。

globally.

The sorts of alien invasive problems spreading in China are many weeds such as the bush Lantana camara, the herbs Eupatorium. Water weeds choke waterways, cause eutrofication of freshwater bodies and kill local fish. Alligator grass is spreading rapidly beside waterways in southern China and huge areas of lakes and slow rivers are now choked by the floating weed of water hyacinth.

Introduced poisonous toads eat and displace more delicate local amphibians. Wood-boring beetles that entered China in the wood of shipping crates but now destroy native forests. Louisiana crayfish, were introduced as a new food item for the Chinese kitchen but are weakening the dykes and flood defenses of the Yangtze floodplains by burrowing deep holes.

Several studies and publications have recently emerged listing and highlighting the damage and threats posed by these new unwelcome species in China. Already more than 300 dangerous species are recognized and the damage caused to the Chinese economy is estimated at $14 billion per year (2004) and growing fast.

Alien crayfish cause damage to the flood dykes of central China but attract ready custom in the street markets.

外来小龙虾在华中地区的防洪堤坝造成损害，但在街头巷市却很能吸引食客。

Invasive climbers smother degraded forest and inhibit reforestation.

入侵性的攀缘植物能窒息已退化的森林，并抑制再造林。

Endemic toads face threats of exotic toads spreading in southern China.

外来蟾蜍在中国南方蔓延，使地方特有的蟾蜍面临威胁。

外来入侵物种

我们必须把外来物种和外来入侵物种区分开来。在几千年的人类历史中，各地互通有无，是自然的，也有重要的经济意义。有的外来物种起到了非常重要的作用，而且由于引进的时间很长，人们已经不再把它们视为外来物种，如红薯、玉米、烟草、芝麻和石榴等。

许多非本土物种被特意引入中国，往往效果特好。这些包括食品，如马铃薯、辣椒、葡萄、番茄、玉米，还有商业品如棕榈油、橡胶、咖啡、木材，用于装点庭院的观赏性植物、观赏鱼等等。在大多数情况下，这种引进是无害的，不会威胁当地生物，但也有引进物品对当地物种引起巨大危害的。

然而，我们习惯于把栖息地的破坏和对野生物种的过度采集看作是对野生物种的最大威胁。现在我们认识到一个新的、更有危害的威胁正逐渐在全球范围内出现，特别是在那些由于人类活动导致自然生态系统发生剧烈变化的地区。那就是来自外来入侵物种的威胁。

简单地说，外来入侵物种是指从自然分布区通过有意或无意的人类活动而被引入、在当地的自然或半自然生态系统中形成了自我再生能力、给当地的生态系统或景观造成明显的损害或影响的物种。

外来入侵物种通常在初次迁移到一个新栖息地时具有竞争优势，因为在新的生态系统中没有疾病或控制该物种的草食动物。在它们的原始栖息地，每个物种都因为疾病或捕食而被控制在一个稳定的水平上，而且疾病或捕食者的密度越大，维持物种平衡的力度也越高。没有这些天然的约束，外来物种能迅速繁殖，直到它的捕食者也进入这个新的生态系统，或这个新的生态系统的其他物种或病原体已经适应取食这一新的食物。

潜在入侵种的蔓延根据人类的贸易和运动呈对数增长。生物体和种子附在冷冻食品上和木质包装箱中；混在粮食或进口的草种子里；夹在国际旅客的衣服和鞋子上；或藏在活植物的土样中被轮船和飞机带到世界各地。有些引进的初衷是好的，但带来的后果却是灾难性的，如出于园艺、农业试验、粮食生产或天然虫害控制目的的物种引进。

中国特别容易受到外来物种的入侵，因为中国有各种类型的的栖息地，可以为进入中国的任何物种提供合适的生长环境。地形的快速变化导致了大片的空旷栖息地，能让外来物种迅速繁殖。中国已经迅速成为世界货物贸易的最大进出口国。

在中国，造成外来物种入侵问题的物种包括很多杂草，如马缨丹和紫茎泽兰。水生杂草会阻塞河道，导致淡水水体富营养化，杀死当地的鱼类。水花生在中国北方的水道和大湖泊旁边迅速生长，水流缓慢的河流如今都被水葫芦阻塞。

引进的毒蟾蜍会捕食和取代更温顺的当地两栖动物。吉丁虫随木板包装箱进入中国，如今正威胁着当地森林。小龙虾被当作新的食物引入中国，但是它善于挖掘深洞，现在正威胁着长江漫滩地带的堤坝和防洪坝。

近年来，中国的一些研究和出版物已经列出和强调了这些新的不受欢迎的物种所带来的破坏。已经确认的危险物种有300多种。2004年，危险物种带来的经济损失估计为140亿美元，且损失增长很快。

The pretty water hyacinth has become one of China's most disastrous alien weeds, blocking waterways and reducing fish stocks.

外观漂亮的凤眼莲堵塞航道，窒息鱼类，成为中国危害最严重的外来植物。

Eupatorium is listed as an alien invasive weed but many farmers like the plant as it covers fallow fields quickly protecting the soil, yet can be easily cut and becomes a good green fertiliser, when a new crop is planted.

紫茎泽兰被列为外来入侵杂草，但许多农民喜欢这一植物，因为它可迅速地覆盖休耕田地，保护土壤。新的作物种植后，紫茎泽兰可以很容易地被割剪下做成绿色肥料。

Genetically Modified Organisms

A new and special threat occurs as China delves deeper into the unknown world of genetically modifying organisms. China already plants extensively genetically engineered varieties of soya bean, cotton, rape, maize, tomatoes, papayas and other cops. The process of such engineering involves isolating genetic material that contains new and desirable genes, inserting the gene to a transgenic vector virus bacterium and introducing the vector into the DNA of the recipient variety.

Although strict controls and tests are in place to ensure that unwanted varieties or genetically unstable forms do not escape into the wild, there is always a risk that this could happen. Wild rape plants (Brassica napus) in Germany, Belgium and Japan already show signs of pollution by genetically modified varieties of rape as a result of accidental spillage of GMO seed destined for use as animal feed. Such plants can easily hybridize with Chinese cabbage via wind dispersed pollen. The danger remains high that man modified organism may spread through wild habitat like alien invasive species or genetically pollute wild and domesticated species, threatening their long-term viability.

Yellow rape flowers add a splash of colour to the countryside.
金灿灿的油菜花为乡村增颜添色。

转基因生物

随着中国对转基因生物的研究不断深入，新的特殊威胁也出现了。中国已经对包括大豆、豌豆、棉花、玉米、土豆、西红柿、番木瓜等在内的多种作物实施了遗传改良。遗传改良工程的过程包括将所期望的基因从含有该基因的材料中分离，将所分离出来的基因注入一个转录媒介，然后将这个转录媒介引入受体的DNA中。

尽管严格控制和检测保证了不让有害的品种或不稳定的遗传形式流入大自然，但也存在随时可能发生的风险。在德国、比利时和日本，野生油菜已经出现被转基因油菜品种污染的迹象，而污染是由于用作动物饲料的转基因油菜种子意外撒落而引起的，这样的植物能轻易地通过风媒授粉与大白菜杂交。人工改良的生物体可能会作为外来入侵物种蔓延到整个野栖息地，或污染野生物种与驯养物种的遗传物质，对它们的长期生存能力造成很大危害。这样的风险是非常高的。

Big face of yellow sunflower provides seeds and oil.

向日葵种子是最受欢迎的消闲食品，也能榨出高档食用油。

Other New Agricultural Practices

Many hundreds of species and thousands of locally adapted varieties of agricultural crops and products have been selected, improved and developed by Chinese farmers over several centuries. The advantages of these produce is that they can be tightly adapted to local conditions of climate, water availability and soil chemistry. They are generally vigorous and disease tolerant. They form the basis of Chinese diet, eating culture and food security. Diversity of forms has its own values. A second variety is always available of conditions become too wet, too hot, too cold for one old favourite. Farmers harvest and replant their own seed and remain in total control of how much and what species they plant on their land. They can switch species as the market profitability of one species or another rises or falls. They grow what they need to feed themselves plus whatever else will bring in the best income.

Modern agriculture advances have hugely increased overall productivity in China. New hybrid crops have higher yields than the original local varieties, mechanized farming has changed the way fields are managed such that small farmers are bought out and larger farming units plant larger fields. Yield are increased by planting under plastic sheets, by use of chemical fertilizers, pesticides and more intensive irrigation systems. New fruits and crops (cotton, coffee, rubber) are planted over the landscape. Farmers plant less for home consumption and more for retail market. They plant less variety but rely on bulk crops and seek the variety in their own kitchen at the market place rather than in the home garden.

The net result of all these changes are that many wonderful varieties representing such a value of selection, local adaptation and disease resistance of crops is being abandoned. Farmers are no longer free to plant what they want. They become tied to planting the seed and varieties being pushed by government programmes or worse—giant multinational agricultural concerns that market infertile hybrid seed.

The loss of variety and so many useful forms is a disaster because it leaves the plant breeders themselves with less building blocks with which to develop and continue to improve current crop varieties. We need maximum diversity because we need to try new combinations, we need to select useful genes from the widest range. We never know where the genes for resistance to a future disease or pests may lie. We do not know what adaptations we will need in face of global warming and fast changing conditions. Rising temperature will certainly cause increases in the numbers of some insect pests. Local varieties may be better adapted to resists such pests but we are throwing them away.

Huge areas of NE China are now planted with valuable ginseng plants but need careful shading from the direct sunshine.

中国东北现已大面积地种植了宝贵的人参，人参须小心遮荫避免阳光直接照晒。

China cannot afford to remain so dependent on irrigation and chemicals. It is clear water resources are becoming ever shorter and our waterways are already becoming dangerously polluted. We need to look for and retain agricultural varieties that offer an adequate even if not maximum yield but offer the trade off of needing less water and less chemicals than some of the new wonder crops being developed.

Preserving seed in a gene bank is only a partial solution. Crop varieties need to be regularly re-exposed to the local conditions to which they are adapted so that the selection process of retaining those individuals that fare best under those conditions can be continued. Without such regular selection the gene pool would revert to the norm from which the variety was originally developed.

Other agricultural changes have negative impacts on biodiversity. Switching from traditional rice fields to planting economic crops such as cotton results in loss of wetland feeding area for wintering cranes. Clearing of land to make way for rubber or bio-fuels causes loss of many bio-rich habitats. Raising of vegetables under miles of plastic sheeting removes from access many invertebrates that birds could eat from the soil as well as leaving a legacy of harmful plastic to dispose of.

新型农业耕作模式

几个世纪以来,中国农民选择、改良和培育了几百种作物和几千个品种。这些产品的优势是:能很好地适应当地气候条件、水分供应和土壤的化学特征。通常,它们生长茁壮,具有耐病性。它们形成了中国食品、饮食文化和粮食安全的基础。这种形式的多样性本身就具有价值。如果种植条件发生了变化,老的品种不再适应太湿、太热或太冷的种植环境,农民总会有备用的品种。农民自己进行采种和种植,自己决定种多少和种什么。他们根据市场和价格的变化,自己消费和创收的需要来种植适宜的品种。

现代农业技术进步已经大大地提高了中国的整体生产力。新的杂交作物比原品种产量更高,机械化耕作已经改变了田地的经营方式,以往一家一户所拥有的小块土地被买断,进行集中耕作。通过使用地膜、化肥、农药和更集约的灌溉系统提高了产量。新的水果和作物(棉花、咖啡、橡胶)被普遍种植。农民为自己消费的种植更少了,更多的是为了市场销售而种植。他们所种植的作物品种很少,更多地种植为数不多的几种大宗作物。为了满足自己对多样食物的消费需求,他们到超市上购买而不是自己种植。

这些变化的最终结果是,许多适应当地条件和具有抗病性的优良的品种被抛弃了。农民再也不能自由选择自己要种的作物了。他们仅限于种植政府提倡的品种,或更糟糕的是,种植大型跨国农业公司推出的杂交种子。

品种及众多有用种类的丧失是一个灾难,因为它留给植物育种者更少可用于开发和继续改良现有作物品种的基础材料。我们需要最大的多样性,因为我们需要尝试新的组合,要在更大的范围内选择有用的基因。我们从来都不知道能抵抗未来病虫害的基因在哪里。我们也不知道为了应对全球变暖和快速变化,需要什么样的适应性。气温的上升将必然导致一些害虫数量的增加。当地品种能更好地抵抗这样的虫害,但是我们却正将它们抛弃。

中国承受不起完全依赖灌溉和化肥农药的负担。事实很清楚,水资源越来越短缺,我们的水道已经被污染到危险的程度。我们需要寻找和保留那些产量不是最高,但比一些新开发的庄稼消耗更少的水和化肥农药的作物品种。

在基因库里保留种子只能解决部分问题。作物品种需要经常暴露在它们所适应的当地环境中,以便在自然选择过程中保留那些适应环境的最佳个体。如果没有这样的定期选择,基因库中保存的不过是陈腐的材料。

其他农业变化也对生物多样性产生不利的影响:从种植传统的水稻转向种植棉花等经济作物导致了越冬鹤类的栖息地的丧失;种植橡胶或生物燃料作物破坏了许多生物丰富的栖息地;由于在塑料薄膜大棚中大规模种植蔬菜,鸟类再也无法从地面上觅食无脊椎动物,因此失去了食物来源;另一方面,还留下了有待处理的有害塑料薄膜。

Agricultural landscapes protect a wide range of agro-biodiversity
农业地貌保护着广泛的农业生物多样性。

Bio-prospecting

With genetic engineering growing as an economic prospect and potentially huge profits available for new wonder crops or wonder drugs, the search is on to find useful genes both within cultivated crops and animals, but also in wild species.

For instance the venom of some biting insects, snakes, scorpions, spiders and marine corals is known to have dramatic effects on heart rate and the nervous system. There are large implications for medical use and development. The insignificant Chinese herb *Artemisia*—so common in semi-desert environment - has been found to have anti-malarial properties and an important and profitable industry can be based on the cultivation and synthesis of this drug.

Plants with disease resistance or with potential for new crops or crop improvement are potentially very valuable.

We call the search for potentially valuable biological material bio-prospecting. Just like other kinds of prospecting this can be done in a controlled and fairly regulated manner under which national sovereignty of its own biological resources is respected and indigenous intellectual property rights of local communities who may know of useful properties or already use certain plants or animals are also respected. But there are also bio-pirates who do not respect these rights and try to illegally collect or smuggle materials out from under the noses of local communities or national governments. It is too late for Brazil to claim rights to the world rubber crop, Zaire to claim copyright of oil palm or China to claim rights to earnings from the kiwi fruit industry. The genetic material has already been exported to foreign lands, developed and refined overseas and now used as a global commodity and trade crop. But it is not too late to tighten up controls and regulations to allow further bio-prospecting; and it is not too late for China to complete its own homework first and undertake a thorough review of its own biological resources so the country knows what it has, where, how much, what it is good for and can then decide whether it wants to grant permits to international, or individual entrepreneurs to make further searches or develop certain materials under trial and eventually commercial basis.

Kiwi fruits originate from China but now a huge export industry of New Zealand.

猕猴桃原产于中国，但现在成了新西兰的大宗出口产品。

生物资源开发

遗传工程在作物改良与新药开发方面创造了一个个奇迹，也使它们成为一个具有巨大潜在利润的经济领域，对有用基因的探索正在进行中，探索的范围既包括现有的作物和动物，也包括野外的物种。

例如，有些昆虫、蛇、蝎子、蜘蛛和珊瑚的毒液对心率和神经系统具有戏剧性的影响，具有重要的医用价值和发展前景。在半沙漠环境中极为普遍的草本植物——黄花蒿被发现具有抗疟疾的功效，培育这种植物进行药物开发，将是一个赢利的产业。

具有抗病性或能用于作物开发或改良的植物有潜在的宝贵价值。

对具有潜在宝贵价值的生物材料的寻找被称为生物开发。如同其他开发一样，生物开发可以在受管制的情况下，以相当规范的方式进行。从而，拥有生物资源的国家的主权能受到尊重，而当地社区可能知道或正在使用某些动植物，他们的知识产权也能受到尊重。但是，也存在不尊重这些权利的所谓"生物资源盗窃者"，他们试图通过非法采集或走私的方式将生物材料秘密地从当地社区或国家运走。

由于生物抢劫发生很早，许多国家已经不能获得对自有生物资源的所有权，例如，巴西不能获得橡胶的所有权，扎伊尔不能得到油棕的所有权，中国不能从世界猕猴桃产业中获得所有权收入。这些遗传物质已经被运到别国，在海外得到开发和提炼，如今已经成为全球使用和贸易的作物。但是，亡羊补牢，未为晚矣。现在加强对生物资源开发进行进一步控制与规范也不迟。对中国而言，首先要摸清自己的家底，对生物资源进行彻底的检查，了解自己有些什么生物资源、在哪里、有多少和有何用途，才能决定是否允许国际机构或个人对某些材料做进一步开发试验或商业性利用。

Divers explore the coral reefs as a source of genetic potential

可以借助潜水设备浏览珊瑚礁。

Too Many Sheep Leads to Too Little Water

Trying to push ecosystem productivity too hard in one direction always results in unwanted reactions in another. Take the grasslands of Qinghai. Many millions of sheep are now grazed there. Mutton tastes nice and the leather can be used but this is not a major contributor to China's economic strength and growth. But trying to raise too many sheep has resulted in degradation of the grasslands. The grass itself has become short and sparse. The hooves of many sheep break up the soil and allow the wind to bite under the vegetation and create deserts and dust storms. Sinking wells to provide water for the sheep lowers local water tables. The raised ground temperatures caused by these processes result in less rain falling, less rain penetrating into the gound, less frost and fog and ultimately lowered water levels in lakes, drying up of small lakes and reduced flow into the Yellow and other important rivers. Downstream communities and industries which really are important for China's economic growth are deprived of the water they need.

A herd of goats can survive in harsh conditions but degrade marginal habitats further and further down the road of desertification.

羊群能在荒芜的的环境条件下生存，但却使本就荒芜的栖息环境进一步荒漠化。

羊多水就缺

物极必反，给生态系统生产力施加了太大的压力，必定会导致相反的效果。青海草场就是一个实例，那里养殖了千百万头绵羊。羊肉美味可口，羊皮可被利用。但是，牧羊对中国经济力量以及经济的增长并没什么贡献。相反，养殖太多的羊导致了草地的退化，草地变得矮小、稀疏。土壤表层被众多的羊蹄破坏后，植被根部遭到强风侵袭，这就形成了沙漠和沙尘暴。太多的羊需要从井中汲太多的水喂养，这样，当地水位就下降了。这些过程导致地表温度上升，降雨量减少。雨水、霜降和雾水的减少最终导致湖泊水位下降，一些小湖泊干涸，也导致了流入黄河及其他一些重要河流中水流量的减少。下游的社区和一些中国经济赖以增长的重要工业也因此丧失了必要的水源。

The Rouergai valley is used by local herdsmen to graze thousands of domestic yak and sheep.

若尔盖被当地农牧民用来放牧数以千计的家养牦牛和绵羊。

Goats huddle by the road side.

山羊在路边挤成一团。

Poaching and Eating Wildlife

In some parts of China, people traditionally eat many wild animal species. In Yunnan you may be served fried wasp larvae or caterpillars taken from hollow bamboo stems. A widely appreciated though now banned dish was the paws of bears. Many restaurants serve turtles, snakes and a range of birds. Millions of sharks are killed each year to take just the dorsal fins for making the famous sharks fin soup.

In southern China there are wildlife markets specializing in live animals—monkeys, civets, squirrels, pangolins, exotic fish and marine creatures—a vast appreciation of biodiversity. But given the huge human population in China and the diminishing habitat and populations of these wild animals, this habit is unsustainable and many of these species are becoming seriously endangered.

China now imports wildlife to meet this demand for food from further and further away. Firstly from neighbouring Myanmar, Laos and Vietnam but now there are wildlife trade networks from Indonesia, Africa and even South America.

Many animals are attributed with dubious medicinal properties in China and this creates a threat to their survival. One recent case in Guangdong involved a gang found to have imported 19 containers of pangolins into China. One container was found to contain more than 2000 of these delightful creatures frozen for sale on the food and medicine trade.

Tiger's penis is considered a cure for sexual dysfunction. Tiger bones for enhancing strength. Even the bones of the tiny zokor—a mole-like rodent considered to

Bag of box terrapins for the food market. China has become a drain on reptile resources of neighbouring South-east Asia.

一袋闭壳龟将被送上食品市场。中国已把东南亚附近的爬虫类动物资源吃光了。

偷猎和食用野生动物

在中国的许多地方，人们传统上回捕食多种野生动物。在云南，你能在餐馆里吃到油炸蜂蛹和竹虫。熊掌曾是广为人们喜爱的菜肴，但现已被禁。许多餐馆供应甲鱼、蛇和多种禽鸟。为了做鱼翅这道名菜，每年有几百万条鲨鱼被杀。

华南有专门的野生动物市场，那里出售猴子、果子狸、松鼠、穿山甲、外来鱼类和海洋动物，包括了多种多样的野生动物。由于中国人口众多，而这些野生动物的栖息地以及种群数量都已萎缩，捕食野生动物不可持续，许多物种已经处于严重濒危的状况。

为满足需求，中国目前从越来越远的地方进口野生动物。以前只从邻近的缅甸、老挝和越南等国进口，现在的野生动物贸易网络已延伸到印度尼西亚、非洲，甚至南美洲。

在中国，许多动物都被认为具某种药用，这对动物的生存就造成了威胁。最近，在广东有一案例，发现犯罪团伙进口了19集装箱的穿山甲，每个箱内就有2000只冷冻过的穿山甲，全部用于出售作为食品或药品。

虎鞭被认为能治疗性功能障碍，虎骨能强身健体。即便是很能挖掘地洞的小鼢鼠（类似鼹鼠种鼠）的骨头也被认为有强健作用，可作药，因而也大量被采猎。

The tokay gecko is now a rare lizard as it is easy to hear its loud calls but it is so prized as a medicinal ingredient.

大壁虎现在非常稀少，原因在于它们叫声响亮，容易被听到，还是珍贵药材。

The Lost Saiga

The saiga is a strange looking antelope with greatly inflated and down-turned nostrils giving it a mule-like profile. The animal certainly has a keen sense of smell but it is believed this organ also serves as a temperature regulator. The stout back-swept horns are heavily annulated and pale in colour.

Migrating herds of saiga formerly numbered in the hundreds of thousands. But sadly due to the belief that the horns have strong medicinal properties, excessive hunting and trading has decimated herds both in China and Russia so that this strange but wonderful creature is now on the verge of extinction. The last Chinese herds occurred in the Dzungarian Basin of NW Xinjiang but the species is now considered extinct in the wild in China. A few animals survive in neighbouring Kazakstan and Mongolia but even here the population is skewed by disproportionate killing of males and breeding rate has declined.

Urgent conservation measures are required to protect the last herds, nurture them back to viable numbers and maybe in the future return the species to some of its farmer haunts including China.

灭绝的高鼻羚羊

Gazelle antlers and other medicinal products in a market.

市场上的瞪羚茸角及其他中药产品。

高鼻羚羊是一种长相奇特的羚羊，鼻孔鼓胀下垂，酷似骡子。它有敏锐的嗅觉，但据说这个器官也起着调节体温的作用。其角粗壮，向后倾斜，有明显的环状纹，呈苍白色。

以前，迁徙的高鼻羚羊群数量达几十万头。高鼻羚羊的角是名贵的中药材——羚羊角，在中国和俄罗斯，高鼻羚羊群已经被过度捕杀。如今，这种奇特的动物正处于灭绝的边缘。中国最后的高鼻羚羊群出现在新疆西北部的准格尔盆地，但目前认为野生的高鼻羚羊已经在中国灭绝。在邻近的哈萨克和蒙古有少量高鼻羚羊存活，但是由于多数雄性高鼻羚羊被杀，繁殖率下降了。

需要尽快采取措施保护这最后的高鼻羚羊群，把它们繁育到具备足够生存能力的数量，并在将来将它们释放回它们原来生活的地方，包括中国。

Smuggling of Hawks and Falcons

The open arid lands of western China are ideal hunting habitat of avian raptors. But of these it is the saker and peregrine falcons that attract unwanted attention from well organized and well equipped smuggling gangs. Even eagles are taken for the falconry trade. These birds command such high prices from falconers of the Middle Eastern countries that operators can afford to have the resources, sponsor the capture, and export of these birds out of the country.

The area is so vast, the boundaries of China so long and rugged and the poachers usually better equipped than national law enforcement teams. So it is easy for poachers to find gaps in the protection network and hard for wildlife protection officers to catch the smugglers who plunder the countries resources of these precious and magnificent birds.

Some consignments get intercepted and the birds returned to freedom but this remains an issue of high concern if these superb hunters are to be preserved. Meanwhile the loss of avian predators continues.

The white-tailed eagle lives along coasts, lakes and rivers and eats fish as well as terrestrial prey. It has the most massive beak of all Chinese eagles.

白尾海隼沿海岸、湖泊和河流生活，也吃鱼类，并在地面猎食。中国的鹰类中它的嘴最大。

走私鹰和隼

　　中国西北地区广阔的干旱土地上生活着很多种猛禽，其中的猎隼和游隼成了组织严密、装备精良的走私团伙觊觎的对象，甚至包括一些鹰类。由于中东国家的养隼者出价很高，走私者有足够的资源来组织捕猎和出口。

　　西北地区幅员辽阔，边境线漫长而崎岖，且偷猎者常常要比执法者装备更精良，因而，偷猎者能轻易找到保护网络的空当而屡屡得手，而野生动物保护者却很难抓获那些盗窃珍禽的走私者。

　　有时，野生动物执法者能在中途截获一些走私的鹰隼，并将它们放归自然。但是，走私猛禽仍然是保护面临的一个严重问题。同时食禽动物的损失也在持续恶化中。

The Saker falcon is the most sought after raptor in western China.

猎隼是中国西部最被猎求的猛禽。

Tourism

Tourism is both a blessing and a threat. It can bring sustainable revenues into poor rural areas which have little other economic production capacity. It can thus relieve the pressure on natural resources and provide an alternative livelihood to communities that would otherwise be forced to exploit their natural surroundings more aggressively.

However, this argument only holds if the substantial revenues from tourism fall to local communities. There is a tendency for these benefits to be gathered and removed by external investors, whether these are entrepreneurs, companies or government agencies. If such external operators try to set up monopolies and exclude local communities from benefiting, the reverse pressure on natural resources can be created.

Then the pressure of large numbers of tourists can cause serious damage and disturbance to natural habitats and species, through noise, pollution, compacting of soils, gulley erosion on trails and roads, and even breaking or collecting of rocks, plants or other components of the natural beauty. Tourism must be kept within the capacity of the given setting and this requires restraint on the part of the tourism operators which conflicts with profits.

In addition, tourism is a fickle industry. Fashions change, new destinations open up and compete with established sites, tourism can collapse in response to natural disaster (earthquake, flood) or security risks. Investments may fail and facilities lay empty.

Some of China's nature reserves attract several million visitors a year which can severely impact the local biota unless carefully controlled.

中国的一些自然保护区，每年吸引数百万游客，如不小心控制，可严重影响当地的生物群。

旅游

　　旅游既是福祉也是威胁。它能为没有其他经济生产能力的贫困山区带来持续的收入,缓解自然资源面临的压力,为当地社区提供新的生计,否则,他们只得以更疯狂的方式利用自然资源。

　　然而,这个观点成立的前提是,当地社区能拿到大部分的旅游收入。现在的趋势是,旅游收益被外来的投资者(企业家、公司或政府机构)占有并带走了。如果这样的外来经营者垄断旅游收益,而当地社区不能受益的话,那么发展旅游反而会增加对自然资源的压力。

　　大量的游客也会带来负面影响,从而对天然栖息地和物种造成严重的破坏与干扰,如喧哗,污染,践踏土壤,修建小道和公路造成的水土流失,破坏和收集岩石、植物、天然景观的其它成分,。旅游必须在现有承载力的范围内开展,这就要求对旅游经营者进行约束,不能让他们为了经济利益而牺牲天然栖息地和物种的安全。

　　此外,旅游是一个风险很高的行业。游客的兴趣会随时间发生变化,新旅游景点不断被开发出来,与已建立的游旅景点争夺游客。旅游业可能因自然灾害(地震、水灾)和安全风险而崩溃。投资可能失败,设施可能被闲置。

Too many tourists are a boon for the local economy but a headache for conservation managers.

游客太多给地方经济带来福音,但却使环保管理人员头疼。

Eco-friendly, zero pollution buses convey tourists around the Jiuzhaigou nature reserve.

绿色环保,零污染的巴士运送游客参观九寨沟自然保护区。

Climate Change

Climate is changing. It is changing faster than predicted. Those changes are accelerated by human activities, especially the cutting or burning of forest and other vegetation and the burning of fossil fuels.

These changes to climate result in overall warmer winters, rising sea levels, more frequent typhoons and other extreme weather events. The changes will add hardship to most human societies, most agricultural systems and most natural ecosystems. It has been estimated that a 2.5 °C rise in global temperature which could happen over the next 30 years would cause extinction of 30% of the world's species.

Although China's per capita emission of CO_2 is not as high as in many developed countries, its huge population means that China is now becoming the single largest emission country in the world.

Climate change in China is already resulting in melting of glaciers which means the loss of water supply to several million dependents. The rate of typhoons hitting southern China has already doubled over the past 30 years causing frequent flood damage. Droughts will also be more frequent, seasons less predictable and vegetation zones will move across the continent as well as up mountain sides. Already new forests are colonizing mountain sides up to 400 metres above the former treeline.

Species are in danger that conditions in their location may no longer be suitable for their survival but they may not easily be able to shift their distribution to where new habitat may now become more suitable for them. Migratory species, following ancient adapted patterns and timings of movement may arrive too early or too late to coincide with preferred conditions for breeding or feeding.

Coastal habitat will become inundated or eroded by wave action of rising sea levels. Coral reefs are becoming killed off by bleaching and acidification.

Climate change poses a big challenge to much of China's biodiversity, but equally biodiversity and the climate amelioration of good vegetation cover can be the quickest and easiest way to mitigate and reduce the effects of climate change for the sake of both human and natural ecosystems.

Glacier on Meili Snow Mountain.

梅里雪山的冰川。

气候变化

气候在不断地发生变化，且比预计的变化要快。人类活动，特别是砍伐或焚烧森林以及其他植被、化石燃料的燃烧，加速了那些变化的发生。

气候变化导致了冬天变暖、海平面上升、台风更频繁，以及极端气候的出现。这些变化将为人类社会、多数农业系统和自然生态系统增加更多的困难。据估计，全球温度在未来的30年里可能会上升2.5℃，这将导致世界上30%的物种灭绝。

尽管中国人均二氧化碳排放量没有许多发达国家的高，但是巨大的人口数量意味着中国现在是世界上排放总量最大的国家。

气候变化已经导致了冰川的融化，这意味着几百万人失去了水的供应。华南的台风数量在过去30年里增加了一倍，造成了更频繁的水灾损失。旱灾也将更频繁，季节变得更难以预测。植被带的分布区域和海拔将发生变化，在有的地区，林线已经升高了400多米。

当生存条件不再适合，而它们又不能轻易地转移到更适于生存的新栖息地时，物种就将处于危险的境地。遵循习惯的线路和时间的迁徙物种可能会过早或过晚到达目的地，从而失去最佳的繁殖与觅食环境。

随着海平面上升，沿海栖息地将被淹没或被海浪侵蚀。珊瑚也会因漂白和酸化而大量死亡。

气候变化对中国的多数生物多样性带来了挑战，但是反过来说，生物多样性和增加合适的植被覆盖也是减缓和降低气候变化影响的最快速、最简捷的方式，从而使人类和自然生态系统受益.

Glacier on Yulong Shan is melting fast.

玉龙雪山的冰川正在迅速融化。

The lakes of the plateau are saline but beautiful

高原的湖泊是咸的，但很美丽。

PART FIVE
Saving China's Biodiversity

Protected Areas

The most determined response to the growing threats to China's biological wealth has been the establishment of a nationwide system of protected areas such as nature reserves, forest parks, ecological reserves and international sites that aim to preserve sustainable examples of all representative ecosystems in China.

China has a very extensive protected area system which has grown fast from small beginnings. In 1988 there were only 300 nature reserves but a steady addition of 100 reserves per year has brought that total up to over 2,500 today covering 15% of the territory in addition to several hundred forest parks and several large ecological reserves. This is well above the global average and IUCN target of 10% coverage. 300 of these reserves are listed as State level nature reserves.

Protected Areas serve multiple functions. They:

- Provide valuable ecosystem services to surrounding and downstream lands

- Conserve populations of important species

- Protect examples of representative ecosystems

- Protect important staging areas for migratory species and breeding sources for species that will disperse and can be sustainably utilised elsewhere

- Protect local culture and connections between local culture and nature

- Allow the natural processes of natural selection and evolution to continue

- Provide sites suitable for development of profitable eco-tourism

- Provide control or opportunities for local sustainable use of natural resources

- Provide opportunities for public education and awareness and living laboratories for continued biological exploration and study

- Protect sources of potentially valuable genetic resources

- Preserve the option of the wilderness experience and recreation to people in a highly urban society.

The State Forestry Administration manages the largest share both in numbers and area but many other agencies also establish and manage nature reserves including the Ministry of Environmental Protection, Ministry of Agriculture, State Oceanography Administration and local governments at various levels.If we examine a map of the distribution of nature reserves, it is clear that there are several huge protected areas

Nature Reserve System in China
中国自然保护区体系

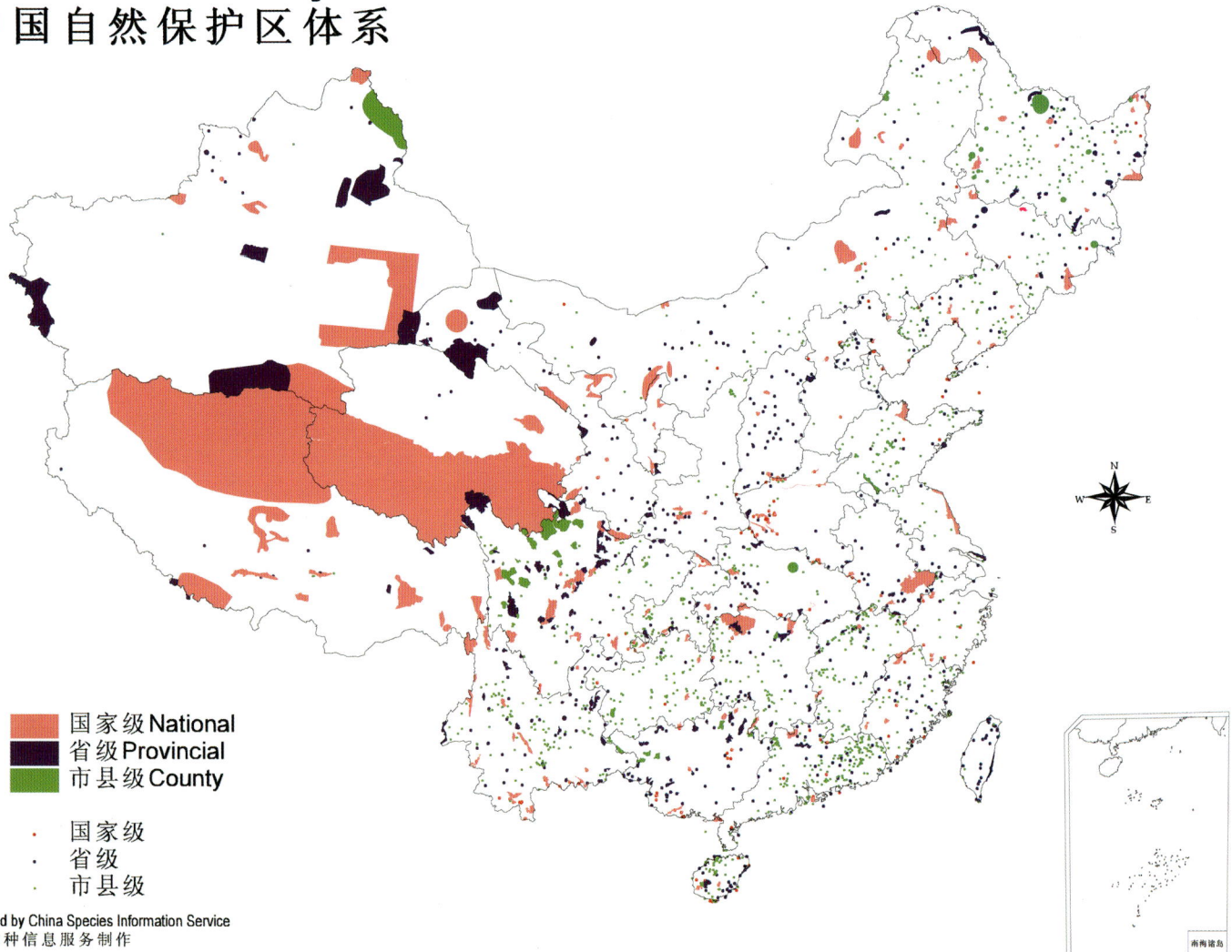

国家级 National
省级 Provincial
市县级 County

· 国家级
· 省级
· 市县级

Producted by China Species Information Service
中国物种信息服务制作

in the west of China and a large number of tiny reserves in the south and east. There are still regions and habitats in China where protected areas are inadequate or too small to protect viable populations of representative local ecosystems and species. There is a need in the east to link these small reserves via corridors of natural or semi-natural habitat so that the reserve system can be linked via such stepping stones and suffer less from isolation. It is important to plan the protected area system more from a landscape perspective than on a site by site basis. It is necessary to plan from an ecological perspective rather than by a species by species approach.

In some places the huge size of the declared reserves gives them an apparent importance they may not merit on biological grounds whilst, at the other extreme, areas of supreme biological importance may not be protected at all, or have only one or two nominal protected areas that are quite insufficient in terms of regional conservation. The situation is also complicated by the fact that there are several different agencies independently in the business of setting up nature reserves, and there is no central agency with complete information on the entire protected areas system.

The pace of creation of new protected areas creates problems of management. Many reserves are inadequately staffed and their staff have no prior experience nor any new training in protected area management. A major programme of training and establishing proper professional standards in what is in effect a new profession is urgently needed. There is also a need to revise and streamline the existing legislation for the establishment, zoning, management, and payment for protected areas and control of tourism within such sites. When protected areas are established without proper funding in place this leads to the development of some contradictory revenue-making activities such as sale of natural resources, creation of zoos, amusement parks and other diversions.

The problem of multiplicity of agencies again arises as it is not easy to get different agencies to pool data, maps and plan together in a holistic manner. Moreover, managing a system of nature reserves is not just a concern of government agencies but should be done in partnership with the general public and private sectors. That beautiful natural areas still occur in China is due to the benign use and management by local communities over the centuries. It is important to retain and foster these conservative aspects and work closely with local communities to continue to preserve such sites.

There are often antagonistic relationships with local farmers feeling that reserves deny them access to lands and resources that their families have enjoyed for hundreds of years. Ways to share benefits with, discuss management options with and enroll the help of local communities in protected area management is important.

第五章
保护中国的生物多样性

自然保护区

为了消除中国生物资源面临的日益增加的威胁，最有效的措施是建立一个由自然保护区、森林公园、生态功能保护区和国际保护地组成的全国保护地域网络，从而把中国所有的典型生态系统的代表置于完整的保护之下.

中国的第一个自然保护区建于 1956 年。一开始自然保护区的发展比较缓慢，到 1988 年，只有 300 个自然保护区，但是此后每年稳定增加 100 个，现在总数已达 2500 个，占国土面积的 15%，远高于全球平均水平与国际自然保护联盟提出的占国土面积 10% 的目标。此外，还有几百个森林公园和几个面积巨大的生态保护区。在这些保护区当中，有 300 个被列为国家级自然保护区。现在中国已经拥有一个种类齐全分布相对合理的保护区系体系。

保护区能发挥多种功能，包括：

• 为周围和下游地区提供宝贵的生态服务

• 保护重要物种的种群

• 保护有代表性的生态系统

• 保护迁徙物种的重要停歇地，以及那些将扩散到异地而被人类合理利用的物种的繁殖地

• 保护当地文化及其与大自然的联系

• 保证自然选择和进化的继续

• 为发展生态旅游提供合适的场所

• 为当地自然资源的可持续利用提供管理和机会

• 为开展公共宣传和环境教育提供机会，为持续的生物利用和研究提供自然场所

• 保护极具潜力的宝贵遗传资源

• 保留城市居民体验野外生活和娱乐的机会

就保护区数量和面积而言，国家林业局管理着绝大部分的保护区，但是许多其他部门也建立和管理自然保护区，其中包括环境保护部、农业部、国家海洋局和各级政府。

如果我们查看一张自然保护区的分布图，能清楚地发现在中国西部有几个面积巨大的保护区，而在华东、华南保护区的数量虽多，但单个面积很小。在中国的很多地区，自然保护区的分布和面积尚不足以保护有代表性的当地生态系统和物种。在华东，要利用天然或半天然的栖息地作为通道，把众多的小保护区联结起来，来减少隔离带来的不利影响。在规划自然保护区体系时，要从景观水平上，生态系统的角度去考虑，而不能就保护区论保护区，就物种论物种。

在中国还存在两种极端情况：一方面，有些地方建立的保护区面积大得惊人，但实际的保护价值非常有限；另一方面，某些极具生物价值的区域可能根本没有得到保护，或只有一个或二个名义上的保护区，并不足以起到区域保护的作用。另外，自然保护区的建立和管理的职责分属于几个部委，它们往往各自为政，而没有一个统一管理全部保护区信息的中央机构，这为自然保护区的有效管理带来了不利后果。

中国自然保护区体系发展过快，也带来了管理上的问题。许多保护区的人员不足，而且缺乏保护区管理的经验，也没受过任何新的培训。因此，对这样的一个新的行业，急需开展必要的培训，并制定相关的专业标准。此外，也要修改、完善关于自然保护区的建立、区划、管理、经费以及保护区内旅游管理的政策和规定。如果保护区没有正当的经费来源，就不得不开展一些与保护宗旨相抵触的创收活动，如出售自然资源、创建动物饲养场和游乐园等。

中国自然保护区管理政出多门，也给数据和资料整合、统一规划带来了困难。再者，自然保护区的管理并不仅仅是政府部门的责任，也需要公众和私营机构的参与。中国之所以还存有些美丽的自然区域，是因为千百年来地方社区对自然资源的良性使用和妥善管理。保护并继承这些传统做法，并与地方社区紧密合作，对保护这些区域至关重要。

另外还需要注意的是，要注意处理好保护区的管理员工与村庄及农村社区的关系。这种关系通常是敌对的，因为当地农民认为保护区剥夺了他们几百年来一直所使用的土地和其他资源。因此，与当地社区共享利益、共商办法和共同管理至关重要。

World Heritage Programme

Some of the most spectacular protected areas enjoy double status in having been nominated and accepted as World Heritage Sites. Most of the World Heritage Sites in China and indeed globally are cultural sites preserving unique examples of human history and culture for future generations to enjoy and appreciate but the World Heritage Programme run by the United Nations Educational, Scientific and Cultural Organisation (UNESCO) also accept nomination of unique Natural sites or Mixed sites which combine natural heritage values with cultural values in the same site.

China currently has 6 natural and 4 mixed sites inscribed on the wall of the UNESCO headquarters in Paris. More sites are in a pipeline in readiness for nomination.

Natural

- Huanglong Scenic and Historic Interest Area (1992)
- Jiuzhaigou Valley Scenic and Historic Interest Area (1992)
- Sichuan Giant Panda Sanctuaries (2006)
- South China Karst (2007)
- Three Parallel Rivers of Yunnan Protected Areas (2003)
- Wulingyuan Scenic and Historic Interest Area (1992)

Mixed

- Mount Emei Scenic Area, including Leshan Giant Buddha Scenic Area (1996)
- Mount Huangshan (1990)
- Mount Taishan (1987)
- Mount Wuyi (1999)

When you have seen the wonderful peaks of Huang Shan, there's no need to visit the other five sacred mountains.

黄山归来不看岳。

世界遗产地

　　一些最壮观的保护区具有自然保护区和世界遗产地的双重地位。中国乃至世界上的世界遗产地大都是文化遗产地，它们保留着人类历史与文化的独特范例，供后代欣赏与品味。但是，联合国教科文组织管理的世界遗产项目也包括自然遗产地，甚至双重遗产地，即同时属于文化遗产地和自然遗产地的地点。

　　中国目前有6个自然遗产地和4个双重遗产地的名字被镌刻在位于巴黎的联合国教科文组织总部的墙上。更多的遗产地已准备就绪，正等待提名。

双重遗产地

- 泰山 (1987)
- 黄山 (1990)
- 峨眉山风景区，包括乐山大佛风景区 (1996)
- 武夷山 (1999)

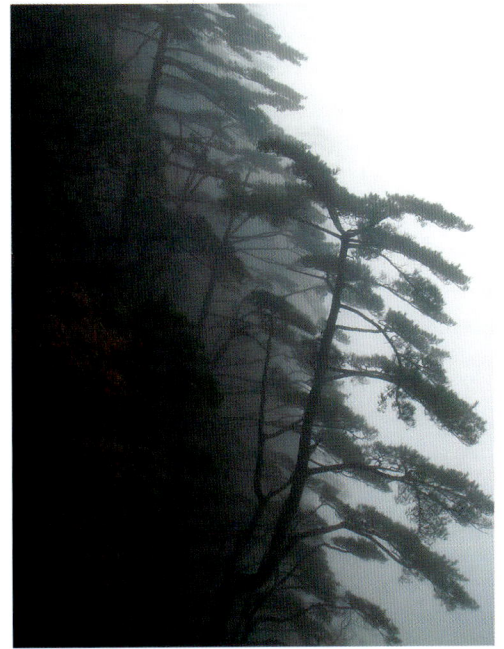

Ancient pines hug the sacred cliffs of Huang Shan

黄山不老松，拥抱神峭壁。

自然遗产地

- 黄龙风景历史名胜区 (1992)
- 九寨沟风景历史名胜区 (1992)
- 武陵源风景历史名胜区 (1992)
- 云南三江并流保护地 (2003)
- 四川大熊猫保护区 (2006)
- 华南喀斯特保护区 (2007)

Strange shaped sandstone pillars give Zhangjiajie its unique atmosphere.

奇形怪状的砂岩柱造就了张家界独特的氛围。

Strange sandstone rock stacks rise above the mists of Zhangjiajie nature reserve.

张家界自然保护区，奇岩怪石从迷雾中崛起。

Colonies of black-crowned night herons live on lakes in many towns in China but range far and wide at night to search for insects and frogs in surrounding paddy fields. The birds utter a harsh croaking 'kowak' as they fly.

大群的夜鹭栖息在中国许多城镇的湖泊，但夜间广布在周边稻田，搜寻昆虫和青蛙。它们边飞边发出粗哑的呱呱叫声。

Macaques are highly social animals. they live in large groups and spend a lot of time grooming each other.

猕猴为群居动物。它们以大数量聚集，花长时间相互清洁美容。

Man and Biosphere Reserves

UNESCO launched its Man and Biosphere programme (MAB) in 1971 and China joined the programme as early as 1973 setting up its own national committee under the Chinese Academy of Sciences. Under this programme member countries establish a system of Biosphere Reserves no both safeguard important sites and to study the international between man and his natural environment. Typically each site has a core area left in a natural condition, a buffer zone where various types of human impact occur and are studies and an experimental zone where much greater levels of human modification and even residence may occur.

The Chinese committee have established 26 such international Biosphere reserves but have also established a further 88 national level biosphere reserves in existing natural reserves across the country.

The following sites are International Biosphere Reserves: As an appendix.

name	Province	Ecosystem
Dalaihu	Inner Mongolia	Lakes, grass and forest
Wudailanchi	Heilongjiang	Wetlands
Fenglin	Heilongjiang	Forest
Xilinguole	Inner Mongolia	Grasslands
Saiwanhula	Jilin	Forest
Changbaishan	Jilin	Forest and mountains
Bogeda	Xinjiang	Tainshan mountains
Juizhaigou	Sichuan	Mountains, forests, lakes
Baishuijiang	Gansu	Forest
Foping	Shaanxi	Forest
Baotianman	Hubei	Forest
Yancheng	Jiangsu	Coastal wetlands
Huanglong	Sichuan	Forest and geomorphology
Shennongjia	Hubei	Forests and mountains
Wolong	Sichuan	Forests, mountains, alpine meadows
Tianmushan	Zhejiang	Forest
Yading	Sichuan	Forests, mountains
Zhumulangma	XizangHimalayan	Mountains
Fanjingshan	Guizhou	Mountains
Wuyishan	Fujian	Forests and mountains
Nanjileidao	Zhejiang	Marine and island
Gaoligongshan	Yunnan	Forested mountain range
Maolan	Guizhou	Karst limestone forest and caves
Xishuangbanna	Yunnan	Tropical forest
Dinghushan	Guangdong	Tropical Forest
Shankou	Guangxi	Mangrove Forest

Protected areas in every country also have some limitations. For example, although important for biodiversity conservation and provision of ecological services, protected areas alone will not ensure preservation of all original species or all the original ecological services. Wild species have large ranges and many now use human modified landscapes as part of their habitat. Isolated small protected areas also lose component species because small populations can lead to fast genetic drift, inbreeding, narrowing of gene pools and demographic catastrophes. Efforts to reconcile high levels of use with high levels of naturalness and lack of disturbance pose obvious conflicts of interest. Poorly planned protected areas may face impossible pressures from local community needs for lands and resources.

The lynx remains a shy and rare hunter of the temperate woodlands.

猞猁羞涩而稀有，在温带林地猎食。

人与生物圈保护区

联合国教科文组织于 1971 年发起人与生物圈计划(MAB)，中国于 1973 年参与了这个计划，1978 年在中国科学院成立了中国人与生物圈国家委员会。根据这个计划，成员国建立了一个生物圈保护区系统，对一些重要的地点进行保护，并开展国际间人与自然的研究。典型的保护区包括一个处于天然状态的核心区；核心区的外围是缓冲区，在缓冲区有各种人类影响，这些人类影响是研究的对象；在实验区有更大的人类干扰，甚至有居民居住。

中国已经建立了 26 个国际生物圈保护区，也建立了 88 个国家级生物圈保护区。

名 称	省	生态系统
达赉湖	内蒙古	湖泊，草地和森林
五大连池	黑龙江	湿地
丰林	黑龙江	森林
锡林郭勒	内蒙古	草地
赛罕乌拉	吉林	森林
长白山	吉林	森林和山脉
博格达	新疆	天山山脉
九寨沟	四川	山脉，森林，湖泊
白水江	甘肃	森林
佛坪	陕西	森林
宝天曼	河南	森林
盐城	江苏	沿海湿地
黄龙	四川	森林和地理形态学
神农架	湖北	森林和山脉
卧龙	四川	森林，山脉，高山草甸
天目山	浙江	森林
亚丁	四川	森林，山脉
珠峰	西藏	喜马拉雅山脉
梵净山	贵州	山脉
武夷山	福建	森林和山脉
南麂列岛	浙江	海洋和岛屿

The argali has massive horns that are butted together in dominance clashes but attract the attention of trophy hunters from all over the world.

盘羊的大型犀角，在占领地的冲突仗时对接在一起打斗，它们也像战利品一样吸引着世界各地的猎人。

高黎贡山	云南	有林山脉
茂兰	贵州	喀斯特森林和洞穴
西双版纳	云南	热带森林
鼎湖山	广东	热带森林
山口	广西	红树林

　　每个国家的保护区都有一定的局限性。例如,尽管保护区对生物多样性的保护和提供生态服务具有重要性,但是单单保护区并不能保证能保护到所有的原生物种或提供原有的生态服务。野生物种的活动范围很大,现在很多野生物种的栖息地保护被人类改造过的地区。孤立的小保护区内的物种更容易灭绝,因为小的种群会导致快速的遗传演变、近亲繁殖、基因资源枯竭和种群灾害。高强度的人类利用与保护自然性和减少干扰之间存在难以调和的矛盾。规划不当的保护区在当地社区对土地和其他资源的巨大压力面前往往一筹莫展。

Man and beast. For many hundreds of years, the gentle buffalo has adapted its behaviour to match the needs of the farmer. Food and care in return for a few weeks hard work. How long will the relationship continue in the face of motorised mini-tractors ?

人兽之间: 数百年来, 柔顺的水牛已适应了农民的需要。它们用几个星期的辛勤工作就能从农民那儿换取食物和照顾。但面对机动小型拖拉机的出现, 这人兽之间的关系能持续多久呢?

Visits to some major nature reserves

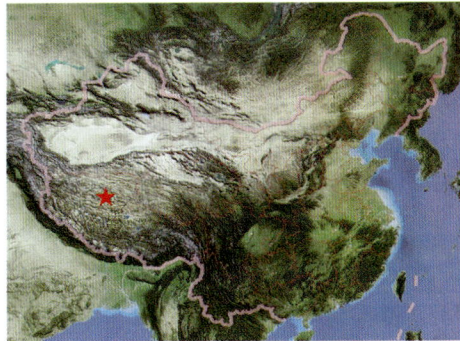

Changtang

Changtang is China's largest protected area in China with a total size of 334,000 km2 or almost the 1.5 times the size of Britain!

The reserve was established in 1993 to help halt the slaughter of wild yak for its meat and chiru for its wool. Chiru wool is the finest known, and is smuggled to Kashmir, India where it is woven into shahtoosh shawls. These shawls are sold worldwide to the wealthy for as much as $15,000 each. Adjoining areas have also been given protection for a current total of about 550,000 square kilometers.

This truly the wildest and most remote place in China, yet great wildlife live in these harsh conditions. In addition to wild yaks and chiru, Changtang is an important area for several rare plateau species including the Tibetan gazelle, kiang, brown bear, Marco Polo sheep, snow leopard, wolf and avian raptors.

Evey year between 4-5,000 female chiru make a 200 km migration north to a desolate but favoured calving area in Xinjiang, then return again in the fall. Numbers have dropped as a reulst of poaching but may now climb as a result of improved protection. Kiang are already more numerous and local herdsmen complain they compete with their herds for grazing.

World Conservation Society (WCS) together with Peking University and local government have been undertaking studies of the wildlife and the needs of local nomadic herdsmen. Under the ECBP programme the project is given further assistance to train guards and find ways to balance the needs of wildlife, rangelands and local communities.

Wild asses trudge slowly through the shallow snow, Changtang in winter.

羌塘深冬，野驴慢步浅雪。

Cheerful herdsman milking the sheep.

欢快的牧民在挤羊奶。

重要自然保护区一览

羌塘

羌塘是中国最大的保护区，面积为334,000平方公里，几乎是英国土地面积的1.5倍。

羌塘保护区建于1993年，目的是保护野牦牛和藏羚羊。藏羚羊绒毛的质量无与伦比，常被走私到印度克什米尔地区，被制成沙图什（shahtoosh）披肩，然后出售给全世界的富人，售价高达每条15,000美元。邻近地区也受到保护，目前保护区的总面积约为550,000平方公里。

这个地区是中国最原始、最偏远的地方，但是许多大型野生动物生长在这些恶劣的条件中。除野牦牛和藏羚羊外，羌塘还是几种稀有高原物种的重要栖息地，其中包括藏原羚、藏野驴、棕熊、盘羊、雪豹、狼和猛禽。

每年，大约有4000到5000只雌藏羚羊往北迁徙200公里，前往新疆的一个荒凉但世世相传的地区产仔，然后在秋季返回。由于偷猎，藏羚羊的数量已经下降，但是加强保护之后，现在数量有所回升。藏野驴的数量已很多，当地牧民抱怨野驴跟他们的家畜竞争牧草。

国际野生生物保护学会、北京大学、当地政府已经对野生动物和当地牧民的需要展开了研究。中欧生物多样性保护项目将对这个项目提供进一步的帮助，包括为保护区培训巡护人员，和寻找协调野生动物、牧场和当地社区发展的办法。

The endemic Tibetan hare perks up its huge ears to listen for signs of danger.

西藏特有的西藏兔，竖起大耳朵，探听有否危险。

Changtang is a vast wilderness. Travel is uncertain.

羌塘原野广阔，旅途茫茫。

Herdsman taking his goods on his domestic yaks

牧民用家养牦牛驮载货物。

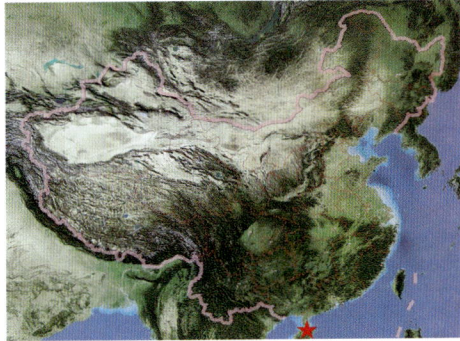

Dongzhaigang

A flock of grey herons rests on the mud banks in front of a backdrop of deep green mangroves. White egrets wade along the sea front, stirring the mud with one foot then suddenly pouncing to stab a small fish or shrimp. Flocks of little plovers scurry among the sand bars chasing insects and small crustaceans. A flash of electric blue and a black-capped kingfisher splashes into the water to emerge with a wriggling catch, fly back to its perch on an old fishing pole and eat its dinner.

Out on the open mudbanks, hundreds of brightly coloured crabs scurry about their business, rolling little balls of sand away from their burrows, feeding like frantic pianists and some fighting away competitors by waving their enlarged right claws. In quiet creeks strange mudskipper fish heave themselves wearily over the mud then lie quietly in wait to snap at a passing sand flea, then rush back to the water to eat the flea with a mouthful of water.

This is the 4000 ha mangrove reserve of Dongzhaigang in the northeast corner of Hainan Island. The reserve is listed as a Ramsar Site and a National level Nature Reserve. It is China's largest natural mangrove reserve, though still bears the signs of having been cut over by charcoal makers in the past. The mangroves are still small but growing steadily. Also the area is used by many fishermen who set their nest across the waterways and rake the low tide sandbars each day in search of mollusks and worms so favoured in the sea food restaurants of Haikou town—the provincial capital, some 27 km distant.

As the tide rises, flocks of birds come closer to shore and eventually fly up to roost in the branches, among the armies of red weaver ants. Fishermen hurry to get their boats back to safe haven and distant rolling sound and darkening sky forecast an approaching thunder storm. Soon the sound of rain with patter on the shiny stiff leaves of the mangroves but the crabs will be tucked inside their burrows beneath the waves.

Small tidal creeks drain through the Dongzhaigang mangroves.

一条小溪在东寨港的红树林流干了。

227

Common Redshank feeds among the tidal mudflats.

红脚鹬在泥潮滩上觅食。

Adventitious rootlets of Avicennia trees mark the leading edge of the mangrove colonisation on new mudflats.

白骨壤林的根系肆无忌惮，标记着红树林侵占新泥滩的最前缘。

东寨港

　　一群灰鹭伫立在海滩上，背后的红树林郁郁葱葱，延展到遥远的天际。白鹭在浅水里逡巡，足带起的泥沙搅浑了海水，长长的嘴巴趁机闪电般地啄住一条条小鱼。成群的小鸻健步如飞，在沙滩上追逐着昆虫和小虾小蟹。犹如一道蓝色闪电，翠鸟如箭般地扎入水中，很快又衔着挣扎着的小鱼浮出了水面，回到了原来的栖身处，尽情地享受着美餐。

　　在空旷的滩涂上，数百只颜色鲜艳的螃蟹正在忙碌着，或将一个个小沙团推出洞外，或四处觅食，十爪移动敏捷，如熟练钢琴家在琴键上翻飞的十指，或挥舞着硕大的右螯向入侵者示威。在一条涓涓细流边，弹涂鱼吃力地爬上泥滩，静静地躺着，突然咬住路过的沙蚕，然后又匆忙跃回水中，就着水享受着美餐。

　　这就是位于海南岛东北角，面积为4000公顷的红树林保护区的景象。这个保护区被列为拉姆萨湿地，是国家级自然保护区。它是中国最大的天然红树林保护区，虽然依旧能看到烧炭者采伐过的痕迹，红树林面积依然很小，但在稳定增长。这里，也有许多渔民在水道里张网捕鱼，或在退潮后的滩涂上用耙子挖掘，寻找蚌类和沙虫。在约27公里处的省会海口，这些动物是海鲜餐馆中的美味佳肴。

　　潮涨时，鸟群不断往高处的海滩走，或飞上红树林的梢头，与大群的红黄肝蚁做伴。渔民匆忙将船停泊在安全的避风港，远方传来了轰隆隆的雷鸣声，天空布满了乌云，预示着一场暴雨即将来临。瞬间，雨点噼噼啪啪敲打着红树林光滑厚重的叶子，而海蟹们则躲进了波浪下的洞穴中。

Salt tolerant flowers live around the mangroves.

耐盐的花卉在红树林周围生存。

Stilted roots of a mangrove Pandanus in Dongzaigang.

东寨港露兜红树林的根高高凸起。

Wintering black-headed gulls use the mangroves as a refuge and feeding area.

越冬红嘴鸥把红树林作为庇护所，并在此进食。

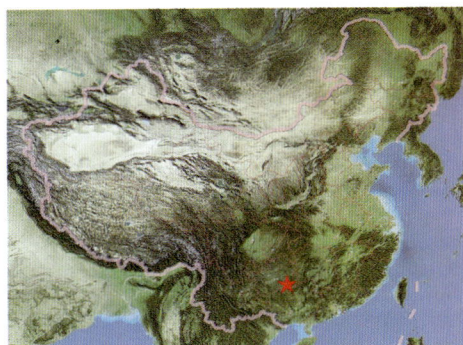

Maolan

It is a steep hot climb among rugged limestone crags - scratched by bamboo and thorny shrubs. In the valley below, farmers are quietly burning the stubble after a successful maize crop. The smell of the smoke hangs in the forest. Butterflies chase among the vegetables and a crying buzzard circles in the sky overhead.

At last we reach the summit of the hill and are surprised to find a series of small rain fed pools in the limestone and even more surprised to find newts and their tadpoles living happily in these pools, feeding on mosquito larvae and other insects.

Colourful spiders stalk unwary flies and jump on them with a sudden pounce. A pink crab scuttles back into a dark recess under a fallen log. Small bitter, cicadas buzz and the black-capped barbet gives its hollow, monotonous humming call. This is the karst limestone nature reserve of Maolan in Guizhou province.

All around stand hundreds of similar limestone peaks forming a landscape of conical hills, each clothed in sub-tropical forest. Each hiding its own little secrets. For the limestone is riddled with cracks and caves, roosts of bats, hiding place for wild serow, nests of silver pheasants.

Karst is a strange rock formation. The main ingredient is limestone which erodes in the heavy rainfall and gets washed gradually away. However, harder types of rock form protected points that serve as peaks when the softer rock around them gets taken away leaving the standing cone stacks, some as high as 300 m.

Plants form an integral part of the landscape. Tree roots both penetrate and split the rocks but then also help bind them together. Rock and plant forms a 3 dimensional living landscape within which myriad millipedes, fungi, orchids, lichens and other epiphytes form a rich ecosystem.

A pleasant wooden visitor centre displays labeled photos of many of the reserve's natural wonders and specimens of the variety of geological rocks and formations that can be found in the reserve. A staircase up an easier hill leads to a platform where visitors can view the strange landscape. A riverine section of the reserve offers rafting in rubber dinghies, views of pretty dragonflies and under a great rock archway, the nests of great flocks of swifts.

The reserve itself totals 20,000 ha, ending at the provincial border. But the limestone hills continue for many kilometers into Guangxi. The home of rare monkeys, snakes, medicinal plants and great caves.

Together with four other sites, including the famous limestone forest of Yunnan—Maolan has been listed as part of the South China Karst World Heritage Site both an honour and a responsibility for effective management.

Even on the steepest rock cliffs, plants are able to gain a hold, tree roots penetrate into the cracks and vines and epiphytes find enough hold and moisture to survive.

即使在陡岩峭壁，植物也能够站住脚，在裂缝中生根发芽，藤蔓和附生植物能生长且吸取足够的水分以生存。

The extensive karst limestone of southern China hides hundreds of km of caves and grottos. Dissolved limestone in the trickling water creates fantastic sculptures in underground chambers - the homes of bats and centipedes.

中国南部广泛的喀斯特石灰岩下隐藏着数百公里的洞穴。溶解了的石灰岩在溶洞室里滴成了神奇的雕塑。而这些洞室也成了蝙蝠和蜈蚣的家。

茂兰

我们大汗淋漓地沿崎岖的石灰岩峭壁攀爬，顾不上躲避路旁竹子锋利的叶缘和荆棘。谷底农民在静静地焚烧着收过了玉米后的秸秆，烟雾袅袅弥漫在树林中。蝴蝶在菜地里翩翩起舞，秃鹰在空中盘旋，时而发出尖叫。

最后，我们到达了山顶，惊奇地发现在岩石表面上散布星星点点的小水洼，水洼里蝾螈和蝌蚪在快活地游弋，觅食着孑孓和其他昆虫。

艳丽的蜘蛛铺开大网，缠住不经意的飞蝇，然后猛然跳过去袭击猎物。一只粉红色的蟹急匆匆地跑回了倒木下黑暗的空隙处。蝉的吱吱鸣声，伴着黑头拟啄木鸟沉闷而单调的叫声。这是贵州茂兰石灰岩自然保护区的景象。

周围耸立着数百座相似的石灰岩峰，形成了一座座圆锥形的小山，构成了一道美丽的风景。每座小山上都布满了亚热带森林，掩盖起各自的秘密。石灰岩上布满了神秘的裂缝和洞穴，成为蝙蝠、鼯羚和白鹇的栖身场所。

喀斯特是一种奇特的岩石结构，其主要成分是石灰岩。石灰岩在大雨中侵蚀，逐渐被冲刷。然而，当松软的石灰岩被冲刷掉以后，更为坚硬的岩石留了下来，形成了耸立的圆锥形石柱，有的高达300米。

植物是这个景观中不可缺少的组成部分。树木根系一方面将岩石穿透和分裂，另一方面也将它们固定住。岩石和植物构成了一个立体的活景观，其中无数的千足虫、真菌、兰花、苔藓和其他附生植物构成了一个丰富的生态系统。

保护区有一个温馨的木屋，那是游客中心，里面陈列着许多带文字解说的图片，展示着保护区的许多天然奇观、岩石和构造。沿阶梯登上并不陡峭的山，从平台上，可欣赏到奇特的景观。在保护区的河流中，可乘橡皮艇漂流，沿河可欣赏美丽的蜻蜓，在巨大的石拱道下，还能看到成千上万的燕子的巢穴。

保护区面积为20，000公顷，到省界为止。但是，石灰岩山延绵几公里，一直延伸进入广西境内。这个保护区是蛇、药用植物和稀有的猴子的家园，也是一个洞穴的集锦。

茂兰与其他四个喀斯特景观一起（包括闻名的云南石林）被列入中国南方喀斯特世界遗产名录，这既是一种荣誉，也意味着一种管理责任。

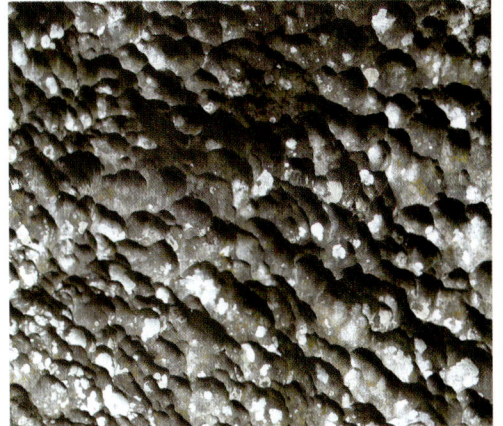

The spray of water condenses on the cave roof then drips from sharp points eroding the limestone into a honeycomb pattern.

溶洞顶上的喷水凝结，然后从顶尖处点滴而下，使石灰岩变成蜂窝格状。

Mysterious mists form over the flooded stream and hang in the windless morning air.

无风的上午，洪水淹没的溪流上面，神秘的迷雾形成了，在空中悬挂。

A small toad lives in the moist limestone forests and depends on the quality of water.

小蟾蜍依赖水的质量，生活在潮湿的石灰岩森林。

Crystal clear water flows off the mountains, purified by the limestone.

清莹的水从山间流出，经石灰岩过滤而纯净。

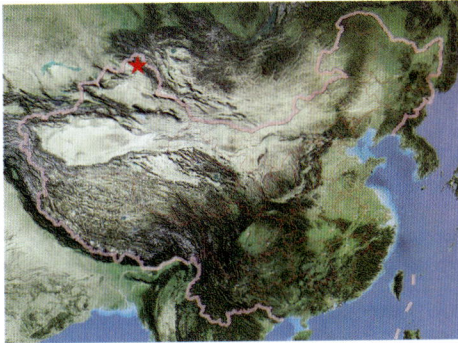

An Introduction to The Shy Lake of Kanas

You heard about her stunning beauty: people say her skin is as smooth as silk, her eyes as blue as the sea. But she is shy, hiding herself deep behind the steep ranges of the Altai mountains, only waiting for those who truly appreciate her to venture in and discover her. So you come, taking the hard route just to have a glimpse of her face.

The winding road leads you to her home. You pass through rugged mountains where tall, straight pines, firs, taigas, and birches pierce into the blue depths of sky. You see deep valleys where white Kazak yurts dot the green meadows like spreading mushrooms, and on the slope sheep grazing, children playing, women cooking, and men carrying poles to build new yurts.

All at once your bus turns around another bend and you catch a flash of a narrow section of the Kanas river, gurgling and bubbling like a cheerful girl; but just as you lean over to the window's edge to see her more clearly, she has already run into the woods again.

You decide to stop and climb the mountains to see her more surely from the above. You follow the cattle path, but soon the rocks, briars and bushes block the path, and you have to cut through the wilderness. You walk between tall pines, cross big logs, carefully avoid stepping on the palaces of ants, and reach a vast patch of pasture and meadow. Oh, what a wonderful view! purple alfalfa and yellow daisies and myriad nameless flowers bloom; happy crickets jump from one blade of grass to another; an eagle circles around chasing or chased by a small bird in the heavens; white clouds float above the peaks like smoke coming out of the woods. You smell greedily the flavor of the fresh dusts and grass; you listen to the concert of insets and hear your own quick breath; you play with your own shadow that has been dragged long on the meadow, or just lay idle under the canopy of birch leaves for a restful moment of day dreaming. It is a paradise with nobody else to interrupt you; only the wooden fence and dried cattle dung remind you that you have intruded into the pasture of Kazak herdsmen.

Everything is so wonderful yet you still feel something missing. Of course, you haven't seen her yet. So after the short break you continue your adventure.

Passing through more woods and crossing over more meadows, you finally, finally reach the peak! And hey, there she is! No longer that giggling girl you met on the side of the road, she is now a charming young lady, gentle, graceful, elegant, dressed up in her emerald-and-sapphire velvet gown, sleeping quietly in the arms of the mountains. Looking at her, you can feel her slow and peaceful

Suddenly the lively Kanas river flows into the quiet and beautiful lake.

生动活泼的喀纳斯河，突然间流入了安静美丽的湖泊。

breathing. What is equally amazing is the background picture stretching out before you, with its own magic combination of colors: the blue sky, green trees, dark mountains, snow-capped peaks, white clouds and that golden setting sun… All you want to do is simply close your eyes, let the wind fondle your hair, feel the warmth of the sunset breathing over you, reflect everything you see in your mind, and you hear your own inner voice whispering: hey, isn't this perfect?

(Text and photos by Liu Jin)

The beautiful swallowtail butterfly lives among the herdsmen meadows, enjoying the same clover plants that Kazak cattle love to eat.

美丽的燕尾蝶生活在牧民的草场上，同样享受着哈萨克族牛爱吃的三叶草。

The Kazak meadows were verdant green with a host of flowers.

哈萨克草场葱葱郁郁，花枝招展。

喀纳斯湖介绍

你早就听说她的美，惊世脱俗——肌肤光滑如丝，眼眸深邃如海。但她的容貌从不轻易示人，羞羞答答深藏在阿尔泰崇山峻岭之间，只待那些有心人跋山涉水前来探访。于是你来了，一路风尘，一路颠簸，为的就是那一眼。

通往喀纳斯的山路曲曲折折。车窗外，掠过座座突兀的高山，山上松、衫、泰加、桦树林层层叠叠，郁郁葱葱，直挺挺，刺楞楞指向苍穹。山间的草场上，白色的哈萨克毡房犹如蘑菇点缀其间，不远处的山坡上，牛羊悠闲吃草，孩童追逐嬉闹，女人烧火做饭，男人肩扛木桩，准备搭建新毡房。

山路盘绕，一转弯，不小心，你撞见了喀纳斯的一道弯，河水汩汩，白浪四溅，一路欢唱有如快乐的小女孩，雀跃向前。你探过身，倚在窗边想看个仔细，哪知害羞的小女孩一转身，又跑进了树丛间。

你不甘心，于是弃车爬山，希望居高临下，一睹她的容貌。你顺马道而行，怎奈半山腰的乱石、荆棘、灌木截了道，你只好披荆斩棘，踏出条进山的道来。穿过高大的松林，跨过倒下的树干，小心翼翼的避开地上耸起的蚁穴，你来到一处开阔的草场边。看啊，风景这边独好！紫苜蓿、黄雏菊，还有无数不知名的野花竞相怒放；快乐的蟋蟀在草丛间跳来跳去；抬头望，一只苍鹰追逐小鸟，在碧空间盘旋斗智；前方的山头，闲云丝丝，有如袅袅轻烟从树丛间升起。你贪婪的深吸一口气，啊，好个泥土野草的幽香。你再侧耳倾听，草丛间虫儿的音乐会开得正热闹，压过你短促的呼吸声。倦了，你玩起自己的影子，任它在草地上拖得老长；或是干脆躺在杉树浓密的树荫下，随随意意做个"白日梦"。如此仙境，无人打扰，唯你独享。只有草间零落的木篱栅和风干了的牛羊粪提醒你，你早已闯进了哈萨克牧人的草场。

一切是如此美好，你却依然若有所失——你还没见到她啊。走吧，歇够了，继续上路。

穿树林，过草场，一路前行，终于，你登上了山顶！看哪，那不是喀纳斯吗？此时的她，不再是那个你在路边碰到的欢呼雀跃的小女孩，而是一位迷人的少妇，高贵、端庄、稳重，裹一袭蓝绿色的天鹅绒长袍，安静的睡在两岸高山的臂膀中；看着她，你仿佛能听到她的呼吸，均匀，自然。你放眼远眺，眼前的图景颜色分明：蓝天、绿树、赭山、积雪压峰、白云悠悠、落日余晖……你闭上眼，任山风吹乱你的头发，感受夕阳的温暖；于是你在脑海中重现这一美景，然后轻轻对自己说：美啊！真美！

（本篇文字及照片由刘津女士提供）

The gentle ripples lap the shore as Kanas becomes a lake.

柔和的涟漪轻拍湖岸，喀纳斯河流变成了湖。

The Kanas River winds like a cobalt ribbon through the emerald forests. This is the last part of China to support Eurasian Beaver and Western Moose.

喀纳斯河, 宛如一条钴蓝丝带穿过翡翠般的森林。这是河狸和驼鹿在中国的最后一片生存地。

High above Kanas, the colourful lichens complete the palette of a dazzling view.

喀纳斯高山上的地衣色彩缤纷, 如调色盘般令人眼花缭乱。

Sacred Forests of Laojunshan

Laojunshan is an area of forested mountains in NW Yunnan. The region is occupied by 7 different ethnic groups including Lisu, Bai, Pumi, Naxi, Yi, Tibetan and Han. Many of the minority groups have sacred forests here for worship or burials within the reserve.

The site is named after Lao Zi the founder of Daoism who used the area to collect medicinal immortality herbs for which the area remains famous to this day.

The scenic area of Laojunshan can be reached by car in two hours from Lijiang town by way of the first bend of the Yangtze river. With assistance of ECBP through its partner TNC, the local township at the entrance to the reserve is set up to cater for tourists in a number of homestays and guides, ponies or even cane chairs are available for those that wish to climb to the 'thousand turtle peak'. The project also encourages a range of other alternate livelihoods such as growing valuable mushrooms, improvement of fruits through grafting and ecological reconstruction works.

Already, on entering the main valley of the scenic area, one is impressed by the great ramparts of sandstone cliffs and pinnacles, the clear streams running out of the forested catchment, the song of birds and the mist rising out of the damp gorges.

The two hour climb to the peak takes you through pine plantations, then mixed oak and conifer forests where village girls scamper to find edible mushrooms, small orchids show their delicate flowers and rhododendron bushes display more flamboyant blooms. In summer the forest is alive with butterflies and the buzzing of cicadas. In winter the chirping of tits and patter of pheasants' feet.

Splendid canyon views greet visitors at every turn and rising tall above the forest is a strange tower that locals call the 'two lovers' but looks rather more like a camel with its two humped hills behind it.

Small begonia flowers and slender arrow bamboos clothe the forest floor till the path turns sharply and the final peaks are in sight. And a strange sight they make as they are weathered into strange 'Buddha head' formations like platted hair of the Buddha statue or as the name of the peak implies it also looks like the knobbly shells of ten thousand turtles.

From the top of the mountain, one can look out for miles and on a good day can see distant snowy peaks. Falcons circle overhead and a distant mooing rises from far below where villagers herd their cows and goats on grassy hillsides outside

The beautiful scene of Laojunshan.

美丽的老君山。

the reserve.

To the west, rise the central ridge of mountains that separate the Yangtze from the Nujiang rivers. Jagged peaks harbour glacier lakes and alpine meadows and sub-alpine forests. Here range troops of the mysterious Yunnan black snub-nosed monkeys. They travel far and formerly connected with monkeys from the Meili Xueshan reserve further north.

On the other side of the ridge the Laojunshan reserve continues and can be accessed from the Nujiang valley road. Here another wondrous site can be visited at the Ninety-nine dragons pools where there are lakes of many colours, streams and waterfalls. Hot water springs allow strange salamanders to live at unusual altitudes.

In total, due to its altitude range up to 4200m, geological diversity, two main valleys and location in the Hengduan Biodiversity Hotspot, Laojunshan offers extraordinary species richness and forms part of the Three Parallel Rivers World Heritage Site.

Brimstone fungus on dead wood.

枯木上金黄色的蘑菇。

Shaped like the shells of many turtles - the Buddha head formations of Laojunshan.

云南老君山的丹霞地貌——千龟峰。

Turtle ? camel ? or two lovers entwined ? Rock formations in Laojunshan.

似龟背？似驼峰？或是情侣相依偎？老君山的岩石。

神圣的老君山森林

老君山位于滇西北的一个山林地区。该区有7个不同的民族，包括傈僳、白、普米、纳西、彝、藏族和汉族。许多民族在保护区内有神山（森林）他们在这里祭祀祖先或墓葬死者。

老君山得名于道家的创始人——老子，他曾在这里采集长生不老的药草，老君山至今仍然以此出名。

从丽江乘车到老君山风景名胜区需两个小时，途经长江第一弯。由中欧项目资助，通过合作伙伴大自然保护协会，地方乡镇在保护区的入口处，建立了旅游宾馆，并给攀登千龟峰的游客提供导游、骑马，甚至滑竿等服务。该项目还鼓励各种替代生计如种植贵重的蘑菇，通过嫁接改进水果品种以及生态重建。

一进入风景名胜区的主要河谷，迎面就是砂岩悬崖像座城墙立即给你留下深刻印象，清澈的溪流从森林集水区冲出，鸟在歌唱，更有薄雾从潮湿的峡谷升起。

两个小时的跋涉攀爬上千龟峰，您途经松林，穿越栎树针叶林混交林，遇上捡蘑菇的小女孩。你看见纤小羞涩的兰花，还有更加艳丽的杜鹃花丛。夏天，森林里有蝴蝶飞舞，蝉嗡嗡鸣叫。冬季，山雀啁啾，雉鸡出没。

每拐过一个山角，壮丽的峡谷就扑面而来，一峰耸立，形似奇塔，当地人称之为"情人峰"，但看上去，更像是一只骆驼的两个驼峰。

小小海棠和纤秀箭竹为森林着装。直到曲幽山径急转，巅峰已经在望。是怎样绝妙的景致！巅峰上裸露的山岩经过风化，形成片片巨鳞状岩块，一座巨大直立的山峰，恰似缠发的辗鹋褴氛。佛陀峰的另一侧，你就看到"千龟竞渡"之景，正如其名：千龟山。

从巅峰极目南望，晴日可看到远处山巅皑皑白雪。雄鹰在空中翱翔，母牛的哞哞叫声从山下远远传来，村民就在保护区外的绿山坡上放牧牛羊。

往西看，蜿蜒的山脊成为长江和怒江的分水岭，崎岖陡峭的峰峦点缀着冰川湖泊、高山草甸和亚高山森林，神秘的滇金丝猴成群在此出没。它们的活动范围很大，在过去与北面梅里雪山保护区的种群联结在一起。

老君山山脊背后仍然属于保护区的范围，可从怒江河谷公路进入。那边风景独好，有另一奇特景观，叫九十九龙湾。九十九龙湾的湖光秀色与溪流瀑布竞争艳。虽然这里的海拔很高，但由于温泉的缘故，仍然有长相奇特的蝾螈生存。

总之，老君山由于高度达四千二百米，地质多样，有两个主要的山谷，位于横断山生物多样性热点，老君山拥有非凡的物种丰富度，成为三江并流世界遗产的一个组成部分。

Mists rise over the hills of Laojunshan.

薄雾从老君山的山峰袅袅升起。

Delicate ground orchid in forests of Laojunshan.

生长在老君山森林地表上的兰花。

Sulfur mushrooms in Laojunshan.

云南老君山的金黄色蘑菇。

The Tianchi crater lake of Changbaishan appears deep blue on a clear day.

长白山的天池，明朗的天，湛蓝的湖。

Changbaishan

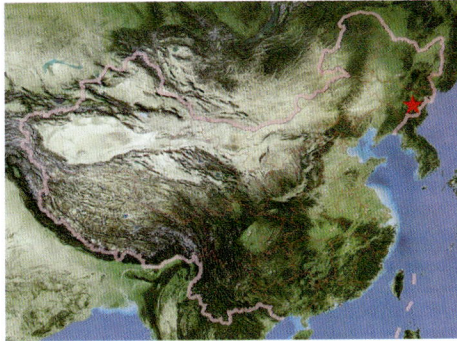

Changbaishan nature reserve is located in the southeast corner of Jilin Province on the border with North Korea. The area was made a reserve as early as 1961 and benefits from extensive inventory work and studies by scientists or the Chinese Academy of Sciences. It is also a Man and the Biosphere reserve under the UNESCO MAB program.

The reserve is in a volcanic zone and the main mountains are composed of volcanic cones and lava flows. The highest peak, Baitu, which reaches 2691m is in fact the highest mountain in North East Asia and is the source of the Songhua, Tumen and Yalu rivers. Mostly within the reserve is China's largest crater Lake - Tianchi, lying at an elevation of 2100m and over 200m deep.

Changbaishan is very rich in flora and fauna. Over 50 mammals have been recorded here, including the rare Siberian tiger, leopard, sika deer, sable, black bear, goral, red deer, lynx and otters. Little chipmunks are common on the ground, inquisitively coming close to visitors been picking their tails and dashing off with the derisory chunk cry. More than 200 species of birds are known from the reserve including such rare species as the beautiful mandarin duck, Oriental stork and Chinese merganser. Changbaishan extends over the border into North Korea, where it is called Paekdu, and would form a good example of an international trans-frontier reserve if both countries would cooperate on its management and protection.

Visitors can rach the reserve both from the south and western entrances. As well as the crater lake, visitors can see dramatic waterfalls, hot water springs, small lakes, a long canyon, open meadows of flowers where the froests were destroyed in the last volcanic eruption and alpine grasslands above the stunted birch and larch forests of the tree-line. Outside the park itself are a number of wildlife farms where you can see tigers and bear.

To the north, the Tianchi Lake drains towards the Songhua plain, cascading down the mountainside in a dramatic waterfall.

天池向北流入松花江平原，在山边形成瀑布，哗啦啦地飘洒而下。

Below and Opposite: Streams flow under the ground for part of their trip from Changbai Mountain northwards into the Songhua plain to eventually join the mighty Heilong river that forms the border with Russia.

下图及下页左图：溪流自长白山往北，部分经由地底下，进入松花江平原，最终加入雄伟的黑龙江，形成与俄罗斯的边界。

长白山

长白山自然保护区位于吉林省东南角，与朝鲜接壤。得益于科学家和中国科学院的广泛调查，这个保护区早在1961年就已经建立。这个保护区被列入了联合国教科文组织的人与生物圈计划。

保护区位于火山区，主要山脉由火山锥和熔岩流构成。最高峰白云峰高达2691米，是东北亚的最高峰，它还是松花江、图门江和鸭绿江的源头。中国最大的火山口湖泊——天池的大部分就位于保护区内，天池的海拔为2100米，深度为200多米。

长白山有非常丰富的动植物，有记录的哺乳动物有50多种，其中包括罕见的东北虎、豹、梅花鹿、紫貂、黑熊、斑羚、马鹿、猞猁和水獭。花鼠在地上几乎随处可见，它们充满好奇地靠近游客，当游客试图去抓它的尾巴时，马上跳开，并发出滑稽的叫声。保护区已知的鸟类有200多种，包括美丽的鸳鸯、东方白鹳和中华秋沙鸭。长白山一直延伸到朝鲜境内，在朝鲜境内被称作白头山。因此，若中朝两国能在管理和保护方面进行合作，长白山将成为国际跨边境保护区的一个范例。

游客可以从南面和西面两个入口进入保护区。在保护区内，除了能参观火山口湖外，还能观看壮观的瀑布、温泉、小湖泊和大峡谷。最后一次火山喷发所毁灭的森林已变成了开阔的草甸，在那里，可以欣赏到鲜艳夺目的各色花朵。分布海拔最高的森林是低矮的白桦和落叶松，再往上就是高山草甸。在保护区外还有许多野生动物养殖场，可以参观虎和鹿。

Clouded yellow butterfly attracted to yellow marigolds.

云黄蝶被黄雏菊吸引。

Cheeky chipmunks are common in the northern forests. They feed in trees and on the ground but usually sleep and nest in a high tree hole.

调皮可爱的花栗鼠常见于北部森林。它们在树上和地面进食，但通常在高树孔里睡眠筑巢。

The common Red Squirrel watches from a safe tree perch.

红松鼠在树上的安全栖息地上观望。

A small lake, a fallen tree, autumn leaves, all is tranquil.

小湖，树桩，红叶，一切是那般宁静。

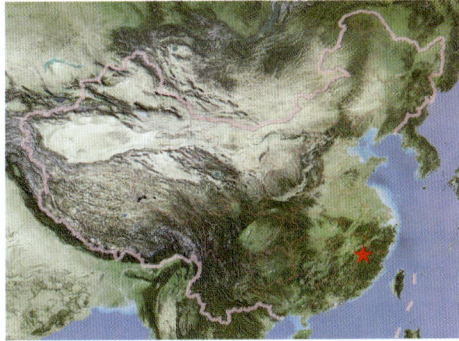

Wuyishan

The World Heritage Site comprises two main natural areas separated from each other by approximately 40 km of roadway. The Scenic area of Wuyishan is the part most visited by tourists and consists of the nine bend river—a stretch of beautiful shallow clear water that twists and turns beneath and between many spectacular sandstone towers and cliffs. The different cliffs are given fanciful names and legends. Indeed some have changed over time. One rock that looks like a giant toad has been recently renamed as the McBurger!

Most visitors float down the river on bamboo rafts but there are land trails across the area with roped paths and stairways up many of the rock piles.

There are prehistoric cliff burial sites, ancient imperial tea gardens and the scholastic temples of the Neo-confucian scholar Lizhi to add cultural interest.

The second portion of the World Heritage Site is the much larger and much wilder Wushishan Nature Reserve which protects some important hill forests and rises up above the treeline to a grassy summit of Huanggangshan at 2158m. The reserve is important for the conservation of many plants and birds and specifically several SE China endemic species such as Cabot's Tragopan and the black muntjac.

The reserve headquarters boasts good accommodation, an interesting museum and troops of tame Tibetan Macaque Monkeys. The fast flowing river is home for the Mournful Kingfisher, lesser Forktail, brown dippers and several species of wagtails and redstarts.

The reserve is buffered to the north by another Wuyishan Nature Reserve belonging to neighbouring Jiangxi Province. One outstanding issue in Wuyishan is the presence of several villages within the reserve and their dependence on cultivating mao bamboo. Extensive hillsides have been converted from mixed forests into almost pure stands of these bamboos. Whilst recognizing that this is a sustainable industry that can meet the livelihood needs of the villagers who live inside the reserve, conservationists are concerned that the bamboo is removing some types of forest within the reserve and forming barriers for the altitudinal migration of many local species. Plans are underway to limit the management of mao bamboo to agreed sectors of the reserve so as to maintain connectivity through that altitude zone for wildlife.

Wuyishan scenic area and the winding nine-bend river.

武夷山风景名胜区，蜿蜒的九曲溪。

Rafts go down the river of 'nine bend stream' in Wuyishan Scenic Reserve of Fujian Province.

福建省武夷山风景区，竹木排沿九曲溪顺水而下。

武夷山

这个世界遗产地包括两个主要自然区，它们相互分离，相距约40公里的路程。武夷山风景区是游客参观最多的区域，其中九曲溪清澈见底，夹岸群峰耸立，峭壁嶙峋。不同的悬崖被赋予了动听的名字与传说。有些名字因时而异，如有一块酷似大蟾蜍的岩石最近被命名为McBurger!

多数游客乘竹筏沿河漂流而下，但是也可以走小道，穿越吊桥，沿梯子爬上石堆。

在保护区内，有史前的架壑船棺，著名的大红袍茶园和理学大师朱熹创建的书院，它们为保护区增添了浓厚的文化气息。

世界遗产地的另一部分是武夷山自然保护区，它的面积更大、更原始，保护着一些重要的山林。武夷山自然保护区的最高峰——黄岗山海拔2158米，已经是在林线之上，植被是草甸。这个保护区对一些物种的保护极其重要，其中包括许多植物，鸟类和几种东南地区特有的物种，例如黄腹角雉和黑麂。

在保护区的总部，食宿服务周到，还可参观有趣的博物馆以及成群的人工驯养的藏酋长猴。保护区河流湍急，常见的鸟类有翠鸟、小燕尾、褐河乌和几种鹡鸰和红尾鸲。

保护区的北面与江西武夷山自然保护区相毗邻。武夷山保护区存在的一个突出问题是保护区内有好几个村庄，村民们靠种植毛竹为生。大片的混交林已经被毛竹林所取代。保护区管理人员一方面认识到，这是一个满足保护区内村民生活需要的一个可持续产业；另一方面又忧虑，因种植竹子，破坏了保护区的一些森林类型，并为许多本地物种的垂直迁徙制造了障碍。目前，保护区管理机构正在计划划定专门的毛竹经营区域，以保证栖息地之间的连接，避免干扰野生动物的垂直迁徙。

Tourists rafting down the pretty landscape of the nine bend river.
游客们从美丽如画的九曲溪漂流而下。

Mother and infant Tibetan Macaques in Wuyishan nature reserve.

藏猕猴母婴在武夷山自然保护区。

Golden tree snake is an agile climber.

黄金树蛇攀爬敏捷。

Jiuzhaigou

Jiuzhaigou is one of the best-known and most beautiful nature reserves in China. It is famous for its series of clear streams, large blue-coloured lakes and many spectacular waterfalls. These wonders are set off against a backdrop of steep forested valleys, sheer rocky cliffs and distant snowy peaks.

In autumn the entire reserve becomes a blaze of rich red, orange and yellow colours as the maples and aspens take on their fall hues.

The reserve has a resident Tibetan population and visitors can enjoy their culture and hospitality in several villages within and outside the reserve borders. Prayer flags blow in the wind. Houses are painted in bright tangka scenes and motifs and on religious days, ceremonies are performed.

The reserve was formerly within the range of the giant panda but pandas have vanished after the heavy flowering of bamboo in the 1980s. The bamboo recovers very slowly but it is hoped that within a few years Jiuzhaigou can once again be a suitable home to this symbolic mammal.

Jiuzhaigou still contains populations of many of the other special fauna of Sichuan highlands such as takin, blue sheep, leopard, musk deer, golden monkeys, eagles and black bears. Bird enthusiast may find sighting the rare black-headed and red-capped robins a more exciting prospect.

Some 30 million visitors come to the reserve each year packed into the summer season and major holidays. The reserve management have adopted a system of eco-buses, closed roads and wooden walkways to limit the spread and impact of the visitors. Most visitors come to the reserve by a,long coach journey from Chengdu, but there is now a Jiuzhaigou airport near the neighbouring and equally spectacular Huanglongshan Nature Reserve.

Tibetans enjoy plenty of space and timber so build large houses and decorate them in style.

藏民喜欢高大宽敞、精雕细琢的木制房屋。

The clarity of water and the strange shades of blue and purple in the calm pools of Jiuzhaigou never fails to impress in any season.

平静的九寨沟池，以其水的清澈和奇特的蓝紫色，在任何季节都留下使人难忘的深刻印象。

Golden rods have become familiar plants in European gardens, but enjoy their real home in the moist woodlands of Jiuzhaigou nature reserve.

金棒兰已成为欧洲花园里的常见植物，但它们真正的家园却在九寨沟自然保护区内的潮湿林地。

九寨沟

九寨沟是蜚声国内外的最美丽的自然保护区之一。它因清澈的溪流、碧蓝的湖泊和壮观的瀑布而闻名遐迩。在陡峭的密林山谷、峻峭的石崖和远方雪峰的衬托之下，这些奇观显得更加迷人。

秋天，枫叶与白杨染上了这个季节特有的色彩，整个保护区成为一个深红、橙和黄色的世界。

保护区内外有几个藏族村庄，游客能享受他们的文化并感受他们的热情好客。经幡在风中摇曳，藏族建筑风格的房屋上绘着鲜艳的唐卡和吉祥的图案。在节日，人们举行隆重的庆祝仪式。

保护区内原先有大熊猫活动，但在上世纪八十年代竹子大规模开花枯死后，大熊猫就不见了踪迹。竹子恢复非常缓慢，但是九寨沟有望在几年后再次成为大熊猫的家园。

九寨沟依然有许多其他特殊动物，例如，牛羚、岩羊、豹、麝、金丝猴、鹰和黑熊。鸟类爱好者可以找到罕见的黑头鸲和红头鸲，需要学名才能确定名称。

每年约有3千万游客来保护区参观，在夏季和主要的节假日，常常是人满为患。保护区管理部门通过使用环保车和关闭公路、铺设木制小道的方式来限制游客的活动范围，减轻对环境的破坏。多数游客乘长途汽车从成都来到九寨沟保护区，有的搭乘飞机，不过九寨沟机场在毗邻的黄龙自然保护区，那里的景色堪与九寨沟相媲美。

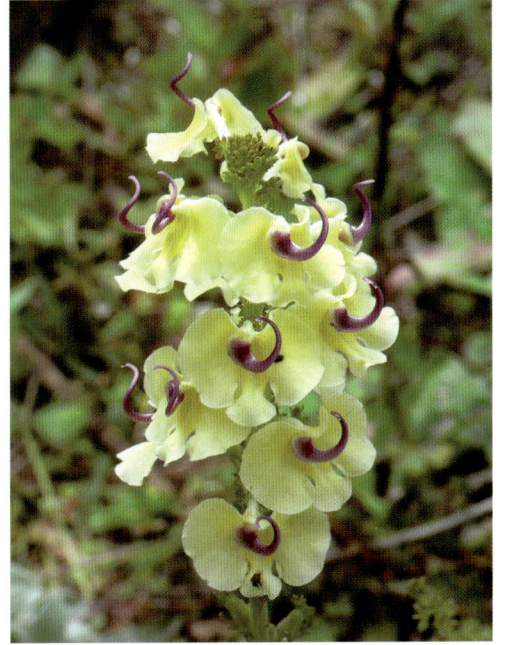

A yellow lousewort among the forest herb layer adds elegance of subtle colour and shape.

森林草本层间的一株黄色马先蒿增添了优雅微妙的颜色和形状。

Like a magic fiord, the long lake of Jiuzhaigou reflects the distant mountains.

就像个有魔力的峡湾，悠长的九寨沟湖映射出遥远的山峦。

Jiuzhaigou's famous waterfalls make it a global beauty spot.

九寨沟著名的瀑布，使它美逾天下。

Red-billed Blue Magpie is a common and noisy bird of mixed forests.

在混合林中经常可以看见红嘴兰鹊这种爱吵闹的鸟。

Xishuangbanna

This nature reserve is famous as the largest tropical forest reserve in China. It is protected as five separated block centred around the prefectural capital of Jinghong, in southwest Yunnan.

Xishuangbanna is home of the richest flora in China. One can see trees of enormous size and at Bubang can explore them high off the ground by way of a suspended canopy walkway.

Xishuangbanna is the only place where visitors can expect to see wild elephants or signs of wild cattle (gaur) in China but also the home of many other fascinating birds, monkeys, reptiles and other wildlife.

The best location to see elephants in China is the Elephant valley at Sanchahe in the Mengyang section of the famous Xishuangbanna National Nature Reserve, about one hour drive from the prefecture capital of Jinghong. Here there are a series of mud wallows and salt licks near a bend in the Sanchahe stream and reserve staff have constructed a series of treetop watchtowers where visitors can view the herds of elephants coming to eat salt rich soil, drink the sweet stream water or frolic in the muddy pools. The treetop observation platforms can be reached via a 2 km walk through the forest along a cement footpath or via a cable car from the main 'Elephant Valley' tourism centre.

Elephants enjoy feeding in the cleared vegetation beneath the cable car, so tourists sometimes get excellent views of the animals browsing directly beneath the overhead cars. Occasionally an elephant will look up or even trumpet at the giggling visitors but more often they carry on feeding and completely ignore this familiar intrusion of their domain.

The forests of elephant valley contain many other tropical specialists that can be found nowhere else in China. These include the rare green peacocks that are adopted as a mascot by the Dai minority of Xishuangbanna, shy wild cattle called gaur, rare monkeys, pheasants, parrots, pythons, 'flying' frogs and other wildlife plus an extraordinary diversity of huge trees and other plants.

The tourist centre contains a large walk through aviary that allows visitors to get close to birds that are shy and difficult to see in the wild forest as well as a large butterfly house where many of the spectacular tropical butterflies of the region are reared and exhibited in a large flight dome for close inspection.

Visitors can also explore ethnic Dai, Hani and Jinuo villages; take part in the famous water-splashing festivals or wander in the large Menglun botanical gardens with its huge collections of tropical plants and interesting museum of ethnobotany.

Blue peacocks have been introduced into Xishuangbanna from India since the native green peacock is almost extinct. The peacock is the prefectural mascot but will young girls now imitate the movements of the wrong species in their traditional peacock dance.

蓝孔雀在西双版纳本土的绿孔雀几乎绝迹后由印度引入。孔雀是西双版纳自治区的吉祥物，但如今少女们在跳传统的孔雀舞时会不会模仿错误孔雀的动作呢？

For those who want more physical challenge there are limestone hills to climb and deep caves to explore. Boat trips from Jinghong can now take tourists down the Lancang Mekong River for a brief view of northern Laos.

Tourists can watch elephants bathing in the river from the Sanchahe watchtowers.

游客们可从三岔河的观象塔上观看大象在河中沐浴。

西双版纳

自然保护区因有中国最大的热带森林而闻名。它由以西双版纳州的州府景洪为中心的五个保护片所构成。

西双版纳是中国最丰富的植物王国。在那里能看到参天大树，在勐腊补蚌架设了一条高20多米、长2.5公里的"空中走廊"，游人可以在上面观赏高大的望天树、原始森林美景和野生动物。

西双版纳不但是中国唯一能看到野象和白肢野牛的地方，而且也是许多迷人的鸟类、猴子、爬行动物和其他野生动物的家园。

在中国观看大象的最好的地方是在西双版纳国家自然保护区勐养片三岔河的野象谷，乘车从景洪到三岔河大约需一个小时。在三岔河溪流的拐弯处有一系列的泥坑和盐窝，保护区员工在许多树顶上建有观象台。在台上，游客能看到象群舔盐泥，喝甘甜的溪水或在泥坑里嬉耍。去观象台有两种方式，一是通过架设在树冠中的长2公里的空中小道，二是从野象谷旅游中心乘缆车。

缆车下的植被被清除得干干净净，大象喜欢在缆车下面觅食，因而，游客有时能在缆车里清楚地看到下面大象吃草的情形。大象偶尔会抬头向上看，或者对着哈哈笑的游客吼叫。但是，更多时候它们只是埋头进食，而对游客置若罔闻，因为它们已经对这种闯入其领地的现象习以为常了。

野象谷的森林包括许多在中国的其他地方不能找到的热带特有物种，包括罕见的绿孔雀（西双版纳傣族的吉祥物）、行踪诡秘的野牛（叫做白肢野牛）、稀有的猴子、雉类、鹦鹉、蟒蛇、"飞行的"青蛙和其他野生动物。另外还有极其多样的乔木和其他植物。

旅游中心有一条通往鸟舍和蝴蝶屋的走道，在那里游客可以走近一些在野外难以看到的鸟类，看到许多当地的五彩缤纷的热带蝴蝶，蝴蝶是人工饲养的，被放进一个大的圆顶饲养室内以供近观。

游客还能参观傣族、哈尼族和基诺族的村庄；参加闻名的泼水节或到勐仑植物园欣赏大量的热带植物和参观有趣的民族植物博物馆。

愿意探险的游客可以选择攀登陡峭的石灰岩山峰，或深入地下洞穴探幽，或乘船沿澜沧江（出国境后叫湄公河）顺流而下，前往老挝北部观光。

At Bubong, visitors can visit the forest canopy by way of long walkways suspended from the tallest trees.

在埠蚌，游客们可从悬挂高树的树冠人行通道观看森林冠层。

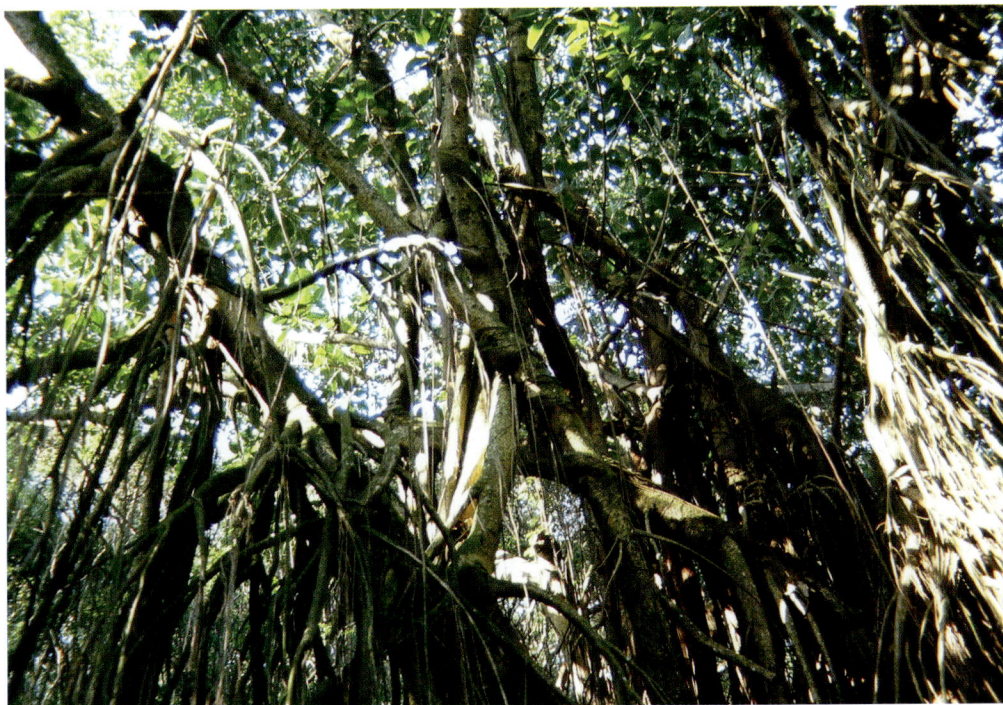

Forest of a single tree- a giant strangling fig tree.

一树成林———一株参天的无花果树

259

A herd of elephants take a mid-day rest in the shady forest.

一群大象在阴暗的森林中午休。

Male fork-tailed sunbird in a Mimosa bush.

叉尾太阳鸟雄鸟在含羞草中。

Poyang Lake

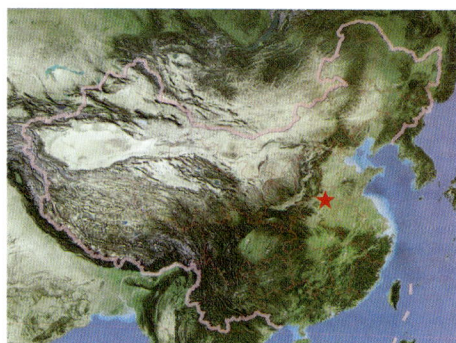

Poyang is a large lake in Jaingxi province in the lower reaches of the Yangtze river but separated from the Yangtze by a short channel beneath the mountain of Lushan. For most of the year the lake is rather small and broken into several separate water bodies fed by three rivers draining south through the lake and into the Yangtze.

In the summer rainy season, however, when the Yangtze is in full flood, the water level in the Yangtze is higher than that in the lakes and water flows in the reverse direction out of the Yangtze and fills the lakes till they all combine to form the full Poyang.

It is the gradual lowering of water level in the autumn which gradually exposed the flooded mudlands and it is this emerging wet habitat that is especially attractive to visiting geese and cranes that migrate south from breeding areas in Russia and NE China to winter at Poyang.

That is the time of greatest activity on the lake, reeds grow and are duly cut by farmers; water buffalo graze, children collect medicinal berries among the scrub bushes, fishermen use nets and trained cormorants to fish in the lakes and eventually set nets over the sluices as the last lakes are drained to hurriedly catch the fattened fish before the next winter sets in.

It is this combination of activities and land-uses that keeps the lake a suitable feeding area for the thousands of birds that winter here, but it is a cycle that may change dramatically once the Three Gorges Dam is complete and the flood waters of the Yangtze are reduced forever.

Most significant bird populations on the lake are the largest global flocks of swan geese, lesser white-fronted geese and almost the entire world population of the endangered Siberian or White Crane. Other cranes, swans geese and a multitude of ducks and other water birds can be seen here. But the lakes are large so you need a strong spotting scope and the winter winds are sharp so you need warm clothing.

Chinese water deer are sometimes seen grazing among the higher sand dunes, hiding up the few scrubby woodlands for much of the day. Otters used to swim in the lakes but are now very rare as are the finless porpoises that sometimes frolic in the larger waterways.

Due to a lack of in-depth study of systematic and precise data, the flood control impacts of Three Gorges Project on the dry season the water level of two major lakes is still controversial. December 25, 2007, the hydrology department of Jiangxi Institute of Science and Technology took the lead in the "Three Gorges Project on the use of Poyang Lake in Jiangxi Province" and the "Five Rivers

Fisherman on Poyang Lake taking his cormorants out to catch fish.

鄱阳湖的渔民带着鸬鹚去捕鱼。

impact" projects. Jiangxi Province has the largest ever water conservancy projects. State financing will allocate special funds of 4.47 million yuan and the Water Resources Department of Jiangxi Province will also be given matching funds to support the project.

The Siberian crane breeds in N Russia but migrates each year to winter in the lakes around Poyang Lake in Jiangxi. Only about 2600 birds remain in the world.

白鹤在俄罗斯北繁殖，但每年冬季都迁徙至江西省鄱阳湖周围越冬。世界上仅有约2600只尚存。

Flocks of geese fill the evening air on a winter night at Poyang Lake.

冬天的鄱阳湖，夜空中有鹤群飞舞。

鄱阳湖

鄱阳湖是江西省的一个大湖泊，位于长江下游，在庐山下的湖口与长江相连。一年中的大部分时间，鄱阳湖水体面积很小，且被分成几个水体。它在北面承纳赣江、抚河、信江、饶河、修河五条大河的河水，向南注入长江。

汛期长江的水位要比鄱阳湖的水位还高，江水倒灌，鄱阳湖水位上升，湖面陡增，水面辽阔；枯期水位下降，洲滩裸露，水流归槽，湖面仅剩几条蜿蜒曲折的水道。具有"枯水一线，洪水一片"的自然景观。

在秋季，水位逐渐下降，被淹没的泥地也逐渐暴露出来，出现一个湿地栖息地，每年吸引着在俄罗斯和中国东北繁殖的水禽和鹤类前来过冬。

这时候也是鄱阳湖人为活动最繁忙的季节：芦苇黄了，农民忙着采割；水牛在草洲上游荡；儿童在沙丘上低矮的灌木丛中采集药用浆果；渔夫或撒网捕鱼，或指挥着鸬鹚在湖里捕鱼；在湖泊慢慢干涸前，渔民将鱼网安装在水闸口，想要在冬天来临前捕捉到肥美的鱼儿。

人类活动和土地利用方式的共同作用，使得鄱阳湖成为成千上万只候鸟理想的越冬地。然而，这种长期以来形成的格局可能会因为三峡水库的修建而发生根本性的改变。三峡水库的蓄水，将在很多程度上降低长江的水位，致使湖水大量泄出，流入长江，从而大幅度降低鄱阳湖的水位，对鄱阳湖的湿地生态系统产生巨大的影响。

在鄱阳湖越冬的有世界上最大的鸿雁和小白额雁种群，世界上几乎所有的白鹤都属于全球濒危的鸟类，以及其他种类的鹤类、水禽和水鸟。但是，由于鄱阳湖的面积很大，在观鸟时需要高倍数的单筒望远镜。冬天寒风凛冽，要注意穿够衣服。

有时能看见獐在突出的沙丘上吃草，但大部分时间它们是隐身在不多的几片灌木林里。过去经常见到水獭在湖里游泳，但是现在已经是难得一见了。和水獭一样，江豚也变得非常少见，人们偶尔才能看见它们在航道里嬉水。

由于目前还缺乏深入的系统研究和确切的数据，因此，三峡工程对于鄱阳湖和洞庭湖这两大湖泊的枯水期水位和防洪格局究竟有何影响，仍有争议。2007年12月25日，由江西省水利科学研究院牵头承担的《三峡工程运用后对鄱阳湖及江西"五河"的影响研究》项目也得以启动。作为江西省水利史上最大的科研项目，国家财政将下拨专项经费447万元，江西省水利厅也将给予配套经费支持。

Young girls pick medicinal berries among the heathers around Poyang Lake.

少女们在鄱阳湖边的石楠丛中采集药用浆果。

The little grebe is a common water bird but not related to waterfowl.

小鸊鷉是一种常见的水上鸟，但与水禽并无关联。

Wolong

Wolong is famous as the home of the giant panda and is also the location of one of the largest captive breeding centres for the giant panda, but it is also one of the largest reserves in central China and protects some of the most valuable biodiversity in China.

Wolong, contains more plants, butterflies, amphibians and birds than most European countries. Five thousand higher plant species have been recorded including more than 50 species of rhododendrons.

Wolong rises through a succession of different vegetation zones from the subtropical broadleaf forests, temperate broadleaf forest zone, mixed conifer forest zone to alpine rhododendron and scrub zone to alpine meadows.

The alpine meadows of Balangshan at 4500m. are the lushest of their type in the world with hundreds of bright coloured flowers all offering themselves for pollination in the short summer season. Several species of red, yellow and blue poppies join the throng of purple louseworts, bright blue Corydalis and plump slipper orchids.

Salamanders hide among the rocks of the mountain streams and snow cocks and blue grandala's hurriedly raise their brood of the year among the rocky screes under the shy shadow of the Siguniang glacier peak at 6250m.

Wolong is home to nine species of pheasant. The most pheasants recorded from any single site in the world. Other bird, mammals and insect taxa are equally well represented and the site is not only rich but also contains a high proportion of endemic species found nowhere else. Key species in Wolong include takin, red panda, blue sheep, white-lipped deer, black bear, leopard, golden moneys and such special plants as dove trees Davidiana involucrata, Duyecao Kingdonia uniflora, Cercidiphillum japonicum and Tetracentron sinensis. The site was one of the seed collecting grounds for the famous British botanist E. Wilson such that many local herbs and shrubs Forsythia, Berberis, Cotoneaster, and the butterfly bush Buddleia are now well-known horticultural plants throughout the temperate regions of the world.

Wolong forms a key site in the larger Hengduan mountains landscape. A region recognized as a global biodiversity hotspot. Together with six other adjacent nature reserves and 9 scenic parks, Wolong has been nominated and approved in 2006 as the World Heritage Site for Sichuan Giant Panda Sanctuaries.

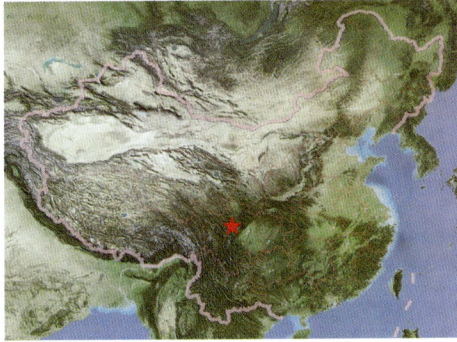

Pitiao valley in autumn in Wolong nature reserve.

卧龙自然保护区皮条谷的秋季。

Giant pandas often climb up trees to rest and get a good view of their surroundings.

大熊猫常常爬到树上休息，还可好好地观察周围的环境。

卧龙

卧龙因为是大熊猫的家园而闻名。它是最大的熊猫人工繁殖中心所在地，也是中国中部最大的保护区之一，保护着一些中国最珍贵的生物多样性。卧龙的植物、蝴蝶、两栖动物和鸟类种类甚至比多数欧洲国家都多。在卧龙，有记录的高等植物有5000种，其中包括50多种杜鹃。

在卧龙分布着从亚热带阔叶林、温带阔叶林、针阔混交林、针叶林、高山杜鹃和灌丛到高山草甸的完整植被垂直分布带。

巴郎山的高山草甸位于海拔4500米处，是世界上同类植被中最茂盛的，有数百种五彩斑斓的花卉，全部在短暂的夏季开花结实。紫色的马先蒿、鲜蓝的紫堇和圆胖的兰花形成了花的海洋，中间还零星的点缀着红、黄、蓝色的罂粟。

蝾螈躲在山间溪流的岩缝中，在海拔6250米、银装素裹的四姑娘山的庇护下，雪鸡和蓝大翅鸲在山坡上的碎石堆中忙着哺育后代。

卧龙有9种雉类，是世界上在单一地点记录雉类最多的地方。其他鸟类、哺乳动物和昆虫均有很好的代表性。卧龙不仅物种丰富，特有物种的比例也很高。卧龙的关键物种包括：牛羚、小熊猫、岩羊、白唇鹿、黑熊、豹、金丝猴和一些特殊植物，如珙桐、独叶草、连香树和水青树。该地曾是英国著名植物学家E. Wilson的采种地，因此，许多当地的草本植物、灌木（连翘、小檗、枸子）和醉鱼草现在成为了世界温带地区众所周知的园艺植物。

卧龙是横断山脉自然资源的精华所在，而横断山脉则是被全球生物多样性的热点地区。卧龙连同附近其他6个自然保护区和9个风景名胜区于2006年以四川大熊猫保护区的名称列入世界遗产名录。

Many species of poppy grow in the mountains. Formerly farmers grew opium in these hills.

许多罂粟花在山上成长。以前农民在这些山丘里种植鸦片。

Wolong's most colourful bird - the cock golden pheasant.

卧龙保护区最艳丽多彩的鸟——红腹锦鸡雄鸟。

Colourful river chat is outdone by the scarlet Sorbus leaves as it feeds on the succulent white berries.

河川有缤纷色彩，但这哪能与鲜红的花楸、汁多味美的白浆果争艳？

Young Tibetan macaques play on a swingy branch and feed on the early buds.

年幼的藏猕猴在摇摇晃晃的树枝上玩耍，吃食幼嫩的花蕾。

Other Conservation Programmes

In addition to the protected area system mentioned above, China has several additional major programmes that also help conserve its valuable biodiversity and preserve the delivery of ecosystem functions of natural habitats.

其他保护项目

除了上述的保护区系统外,中国还有其他有助于保护珍贵的生物多样性和保持自然栖息地供应生态服务的大型项目。

Ex situ conservation

Chinese technicians have a firm belief in the effectiveness of captive breeding programmes and a valid method for recovering diminished wild populations and repopulating wild population. World-wide such efforts have been fraught with problems, faced many failures and even when successful have raised concerns about the sustainability and the ability of captive-born animals to ever become really wild. Wild rescue and captive breeding is generally only practiced as a last resort when in-situ conservation seems doomed to fail.

Many of the efforts to breed rare animals in China may well be motivated out of economic interests and most of the rare pheasants, monkeys and other wildlife in China's more than 300 breeding farms and zoos will probably never result in returning captive bred animals back to the wild. Nevertheless, out of this experience has grown a considerable competence and in several cases China has shown the world that it is indeed possible to restore an almost extinct species to large enough numbers that re-introduction is possible.

移地保护

中国的技术员坚信,通过有效的人工繁殖项目和恰当的方法,能恢复日趋减少的野生种群,并为野生种群增加新成员。世界范围内的这种努力被许多问题困扰,面临许多失败,就算繁殖成功,还要考虑人工繁殖的动物能否适应野外生活。只有在就地保护注定要失败时,才会考虑把野外拯救和人工繁殖作为最后的手段。

The results of a successful breeding programme. Crested Ibis was saved from extinction and now several hundred are returned to their habitats in southern Shaanxi Province.

因为一个成功的繁殖项目,朱鹮免遭灭绝,至今有几百只鸟回到了它们的栖息地——陕西南部。

在中国，繁殖稀有动物的动机可能是为了经济利益。在全国300多个养殖场和动物园内，把人工繁殖的稀有雉类、灵长类和其他野生动物放归野外的可能性非常小。然而，实践出真知，中国已经向世界证明，她成功地把数种濒临灭绝的物种种群恢复到可以考虑再引进的水平。

Pere David Deer - Milu

The Milu—a tall spectacular deer - once roamed in the marshy coastal plains of NE China and a captive herd was maintained at the imperial hunting grounds of Nan Haizi just south of Beijing. It was here that in 1865 the young French missionary Pere Armand David spied the animals by climbing a hill and peering over the hunting ground walls. By bribing the keepers he was able to smuggle out some antlers and fur and immediately realised that this was a new species. He sent the materials back to France where the deer was described and named after him. A few years later, using diplomatic channels he was able to acquire a few specimens to export back to Europe. It is just as well because during the Boxer rebellion of 1900-1 the hunting ground was destroyed and the deer killed or escaped. They became extinct in China but have thrived in Europe.

In 1980s two initiatives—one led by the Duke of Bedford and another by WWF set about rei-introducing captive bred milu to China. Holding farms for these returned herds were established at the original hunting grounds site of Nan Haizi and also at Defeng Nature Reserve on the East coast of Jiangsu Province.

Both herds prospered and now number many hundreds of animals. Both projects have already released some of their animals back into the wild. So once again captive milu roam the wetlands of the east coast and also the middle Yangtze valley. Whether they survive or die out will depend on controlling hunting around them wherever they may wander. And wander they will because these are tall deer than can jump a fence or swim a river with ease.

麋鹿

麋鹿高大华丽，以前出现于中国东北平原沿海的沼泽地，但在北京南郊的南海子皇家猎苑有一个人工饲养种群。1890年，年轻的法国传教士Pere Armand David 爬上一座小山，隔着猎场围墙，偷窥苑内的动物。通过贿赂管理员，他获得了一些鹿角和毛皮，马上意识到这是一个新物种。他将材料送回了法国，并以他的名字对鹿命名。几年后，他通过外交渠道获得了几只活的麋鹿，并将它们运回了法国。1900年永定河发洪水，冲垮了猎苑的围墙，很多麋鹿逃出了猎苑，被饥饿的灾民捕食。同年，义和团事件爆发，猎苑被毁，麋鹿也销声匿迹。后来在中国灭绝，但在欧洲存活了下来。

二十世纪八十年代，贝德福德公爵和世界自然基金开始从欧洲将人工繁殖的麋鹿引入中国，在原先的南海子猎苑和江苏省东海岸的大丰自然保护区饲养。

两个地点的鹿群都增加了，数量有好几百头。两个项目已经将部分麋鹿放回自然。九十年代初，麋鹿又被引进长江中游的湖北石首市的天鹅洲自然保护区，也被成功地放归野外。现在，人工养殖的麋鹿再次徜徉在东海岸和长江流域中游的湿地。它们在野外的命运，完全取决于是否能控制人为猎杀活动。它们能随意到达所想去的地方，因为它们如此高大，能轻易越过围栏和游过河流。

The stag milu or Pere David's deer has a magnificent rack of antlers
麋鹿的角雄伟壮观。

Giant Panda

In the mid 1970s and again in the mid 1980s the bamboo that formed the staple diet of the rare giant panda flowered, set new seed and then died over wide areas. Many pandas starved, some traveled over mountains and across farmed valleys to seek not yet flowered bamboo groves. About 100 animals were brought into captive zoos. The wild population was estimated to number less than 2000 animals and captive breeding was determined to be an integral part of a recovery plan.

Although famously difficult to breed in captivity, Beijing zoo had managed to breed pandas by both natural and artificial methods. In 1981 a new breeding centre was built in the Wolong Nature Reserve of Sichuan by the then Ministry of Forestry with funding and technical assistance from WWF and Beijing zoo. A year later Wolong succeeded in breeding its first panda in 1987. In 1988 another captive breeding facility was opened outside Chengdu. Most of the captive pandas being held at various nature reserves and zoos were transferred to these two centres and gradually breeding success has improved and a crop of 20-30 new babies get born each year.

Experts disagree on how to try re-introducing such captive born animals to the wild and one early re-lease in Wolong proved tragic when the released animal died from injuries probably inflicted by wild pandas not willing to accept this new-comer. Trials may continue in several localities where pandas used to occur, bamboo condition is good enough but there are not or only very few wild pandas.

All efforts are currently on hold whilst the effects of the disastrous earthquake of May 2008 in the wild panda stronghold of the northern Qionglai Mts. is assessed. Wolong itself was badly damaged.

大熊猫

在上世纪70年代中期和80年代中期，大熊猫的食用竹大规模开花枯死，造成许多大熊猫因饥饿而死，一些大熊猫翻越山岭，穿过有农田的山谷，寻找未开花的竹林。大约有100头大熊猫被"抢救"进了动物园。估计大熊猫的野生种群数量不到2000头，人工繁殖被确定为大熊猫恢复计划中不可分割的一部分。

尽管人工繁殖极其困难，北京动物园还是设法通过天然交配和人工授精两种方法，成功地繁殖大熊猫。1981年，由世界自然基金会和北京动物园提供资金和技术支持，林业部在四川卧龙自然保护区建立了一个新的繁殖中心。1987年，卧龙成功地繁殖了第一只大熊猫。1988年，在成都郊外又建立了一个大熊猫养殖场。保留在不同保护区和动物园的多数人工繁殖的熊猫被集中到这两个繁殖中心。渐渐的，繁殖的成功率提高了，每年能繁殖出20~30只熊猫幼仔。

如何将人工繁殖的大熊猫放归野外，专家们尚未达成一致意见。已经进行的一次野放没有成功，可能是释放的大熊猫在与野生大熊猫的争斗中受伤后死亡。野放试验可能还在几处继续进行，释放地点的选择标准是：过去曾经有熊猫分布；竹子资源充足；目前没有野生熊猫或数量很少。

2008年5月，四川发生地震，卧龙保护区遭到严重破坏。在评估地震对邛崃山北部的野熊猫生存地的影响期间，所有的野放试验都暂停了。

Captive bred alligator ready for release in the wild.

圈养繁殖的扬子鳄，准备释放回野生环境。

Lantian born to mother Lili in 1987 was the first captive panda to be born at the now famous Wolong panda breeding centre.

1987年在著名的卧龙熊猫繁殖中心由母亲莉莉生出的蓝田，是第一只圈养出生的大熊猫。

Yangtze Alligator

In the 1960,s there were estimated to be 30,000 wild alligators in the wetlands and lakes of the lower Yangtze. Today the estimate is less than 100. It is on the verge of extinction.

But Chinese conservationists have reproduced more than 10,000 Yangtze alligators through artificial means in recent years. The Anhui Province alligator reproduction center, the largest of its kind in the world, houses 9,000 alligators. The question is now 'What to do with them?'

Certainly many will be released back into the wild but is the wild habitat still suitable for them? Most of the lakes are now isolated from the main rivers. Most waterways are heavily polluted and over-fished. The human population in this area is very dense and will the population be happy to live with wild alligators back in the rivers and lakes? WCS have been fitting radio collars to wild bred alligators and releasing them at selected sites to see how they fare. Clearly there is the technology to maintain the species, but can we find enough undisturbed habitat where they can thrive once more in the wild?

扬子鳄

上世纪六十年代，在长江下游的湿地和湖泊大约有 30，000 条野生扬子鳄，今天只剩下不到 100 条，已处于灭绝的边缘。但是，近年来，通过人工方法已经繁殖了 10，000 多条扬子鳄。安徽省扬子鳄繁殖中心是世界上最大的鳄鱼繁殖中心，目前饲养着 9，000 条扬子鳄。现在的问题是如何处理它们？

当然，许多扬子鳄将被放生到野外，但是是否野外的栖息地能适合它们呢？大多数的湖泊如今已与大的河流割断了联系。多数水道被严重污染和捕鱼过度。这个地区人口稠密，人们愿意在河流与湖泊中再见到野生的扬子鳄吗？野生动物保护学会将无线电项圈套在野生扬子鳄身上，然后将它在选定地点释放，观察它的活动。现有的扬子鳄人工繁殖技术非成熟，维持这个物种的生存不存在问题。现在的问题是，能找到足够的没有受到干扰的栖息地，恢复扬子鳄的野外种群吗？

Crested Ibis

These beautiful bird formerly bred in wetlands of NE China, Eastern Russia and Japan with non-breeding birds visiting Korea and Taiwan. But disturbance and persecution throughout its range was thought to have led to extinction until a small population of only 7 birds was discovered in Shaanxi Province in 1981. Since that time, through a combination of protecting the last wild birds together with captive breeding in two main centres, the species has been save. By 2002, the wild population was maintaining a steady increase and numbered 140 birds, and the captive population was up to 130 birds. By 2006 the population was estimated at over 500 individuals Breeding success has been high in the wild and in captivity. There are new plans to begin reintroduction of birds to Sado Island, Japan.

May 31, 2007, China conducted its first experiment in the reintroduction to the wild of captive bred Crested Ibis. 26 birds were released in the depths of the Qinling Mountains of Shaanxi Province in the village of Zhai Gou, Ningshan County. After nearly a year of field observation, 6 of the birds flew back to the original base, 5 have been confirmed dead, 3 missing, and the remaining 12 survived a rare snowfall last winter, showing they had adapted to the survival in the wild environment and becoming the first captive bred ibises to return successfully to the wild. One pair of crested ibis has been successfully breeding in the wild and produced three chicks.

In the future, China will also release captive bred crested ibises in Shaanxi, Henan, Hubei and other places to gradually recover the wild populations to its historical range. At the same time, other countries in the historical rangey of Crested Ibis will also provide restoration of the wild population.

朱鹮

以前，美丽的朱鹮在中国东北、俄罗斯东部和日本的湿地进行繁殖。在非繁殖期，也常飞到朝鲜和台湾。由于整个生存地受到干扰和破坏，人们曾一度认为朱鹮已经灭绝。后来在 1981 年，中国科学院动物研究所的刘荫增在陕西省洋县发现了一个由 7 只朱鹮组成的小种群。从那时起，通过野外保护与人工繁殖相结合的方法，朱鹮被拯救了。到 2002 年，野生朱鹮的数量保持稳定增长，已达 140 只，人工养殖的朱鹮达 130 只。2006 年，朱鹮的数量估计已超过 500 只。野外和人工繁殖的成功率都很高。有计划将朱鹮引入日本的佐渡岛。

2007 年 5 月 31 日，中国进行了首次朱鹮再引进试验，26 只人工繁殖的朱鹮被释放在秦岭深处的陕西省宁陕县城关镇寨沟村。经过近一年时间的观察，野化放飞的 26 只人工繁育朱鹮中有 6 只回到了原来的放飞基地，5 只已经确认死亡、3 只失踪，其余 12 只经历了去年冬天罕见大雪的考验后，目前已经适应了野外的生存环境，成为第一个由人工饲养变为野外独立生存的朱鹮家族。其中的一对朱鹮已经成功地在野外繁殖出了 3 只幼鸟。

今后，中国还将在陕西、河南、湖北等地对人工朱鹮野化后进行异地野外放飞，逐步恢复野外种群的历史分布区，同时为其他国家朱鹮历史分布区恢复野外种群提供经验。

Crested Ibis has been saved from extinction through captive breeding and is now being successfully reintroduced into the wild.

朱鹮幸免了灭绝，圈养繁殖的鸟现已成功地被引入到野生种群。

Crested Ibis flying free in Shaanxi Province.

朱鹮在陕西省自由翱翔。

Logging Ban

Following disastrous and particularly sever floods in 1998 which displaced millions of citizens and caused damages to industry and agriculture estimated at $36 billion, Chinese prime minister Zhu Rongji ordered a logging ban covering 100 million hectares of the middle and upper reaches of the Yangtze and Yellow river valleys. The ban remains in place to this day and its scope extended being applied in 18 provinces and by greatly reducing the timber harvest quotas for other regions such as NE China and Inner Mongolia. Several hundred timber production companies were closed down and redirected into reforestation and forest protection efforts. China in turn has had to look outwards to meet its ever growing demand for construction timber and has now become the world's single largest importer of timber, although much of this is worked into furnishings for the re-export market.

The tall bole of Parashorea chinensis in the rainforests of Bubong, Xishuangbanna.

西双版纳埠蚌雨林望天树的高树干。

天然林保护工程

　　1998年，中国发生了灾难性的大水灾，导致了数百万居民流离失所，工农业损失达360亿美元，国务院总理朱镕基下令禁伐长江和黄河流域中上游的一亿多公顷区域内的天然林。今天，天然林禁伐依然有效，且范围扩展到了18个省，并对一些重点林区，例如东北地区和内蒙古，降低了采伐配额。有数几百家森工企业被关闭，转向造林和护林。结果，中国必须依赖进口木材来满足日益增长的需求，现已成为了世界上最大的木材进口国。当然，有很大一部分进口木材被制成家具后，又用于出口。

Re-greening is now a huge industry in China but preference should be given to mixed plantings of fruit and flower-bearing local species that can be used by local wildlife.

重新绿化现为中国一巨大产业，但应优先考虑混合种植能受惠于本地野生动物的地方水果和花卉品种。

Reforestation Programmes

China is globally by far the most active country in reforestation, planting 4 million hectares of new forests each year—probably more than the rest of the world combined. It is targeted to restore forest cover over 20% of the land surface by 2010. Several different programmes contribute to this total including programmes to protect remaining forests, restore forests on bare lands, return forest and grass to steep farmlands, build a green wall to combat desertification and major loans for reforestation under World Bank support. These efforts to re-green the countryside are matched by programmes to plant trees and create green spaces in urban areas also.

造林工程

中国是目前全球造林最积极的国家，每年造 400 万公顷新森林，可能比世界其它国家造林面积的总和还多。中国的目标是到 2010 年将森林覆盖率恢复到国土面积的 20%。为了实现这个目标，实施了几个不同的项目，包括保护现有的天然林，荒山造林，退耕还林，防护林建设和世行贷款造林项目。除了这些项目外，还有城区植树和绿化项目。

Pawlonia is the worlds fastest growing temperate broadleaf tree. It holds great potential for reforestation.

Pawlonia 是世界上生长速度最快的温带阔叶树。它蕴藏着巨大的再造林潜力。

Combating Desertification

The deserts of northern China have advanced 300km in the last 3000 years as a result of overgrazing, agriculture and deforestation. But the pace is accelerating and now the advance is an estimated 2,500 km2 per year.

More than 12 billion Yuan have been spent in the past decade on combating desertification in northern China but the damage caused by desertification is estimated at $2-3 billion per year.

The Great Green Wall is a project to plant a 4,480km shelterbelt of trees across the northwest rim of China skirting the Gobi Desert. This is a massive undertaking ans reflects the concern that the government is placing on the creeping advance of the desert and the frequency of dust storms affecting northern cities such as Beijing.

荒漠化防治

Experiments in combating desertification at Hulunbier.

呼伦贝尔在进行抗荒漠化实验中。

由于过度放牧、农业活动和毁林，中国北方的沙漠在过去的3000年里推进了300公里。但是现在前进的速度正在加速，估计现在每年新增加的荒漠化面积达2500平方公里。

在过去的10年里，已经花了120亿元人民币用于中国北方防治沙漠化，但是沙漠化所带来的损失估计为每年20~30亿美元。

"三北"防护林工程计划在中国西北沿戈壁沙漠边缘种植4480公里的防护林带。这是一个巨大的工程，反映出政府对沙漠的蔓延和沙尘暴频繁袭击北方城市包括北京的关注。

"Three North" Shelterbelt Construction

In 1978, the Chinese government decided to build the Three-North Shelterbelt System. Three North region covers all of northern China from NW Xinjiang to NE Heilongjiang—a length of 4480 km and with a width of between 560 to 1,460 km from north to south. The region covers 13 provinces 551 counties, with a total area of 406.9 million hectares, accounting for 42.4 percent of China's total area.

The region was formerly rich in forests and grasslands but after wars and population surge large areas of forest and grassland had been destroyed. By the mid-1970s, the region's average forest coverage was only 5%, the region was undergoing fast desertification with regular huge sandstorms, low rainfall and rapid levels of soil erosion.

Three-North Shelterbelt construction project is designed to last 73 years (1978 to 2050), in three phases: Phase 1 plans to use 23 years reforestation of 21.774 million hectares, the construction of a number of regional shelter forest system; second stage of 20 years reforestation of 8.017 million hectares, forming the basic shelter forest system; and a third phase of 30 years reforestation of 5.233 million hectares, is mainly a battle to consolidation and improve the entire system. By 2050, the Three North Forest area will be increased to 60.57 million hectares or 15 percent, forest reserves increased 4.3 billion cubic meters, ecological and economic benefits accumulated up to over 13,000 billion RMB.

From it launch in 1978 to 2004, the Three-North Shelterbelt Project has already completed a total of 23.5 million ha afforestation, the largest reforestation project in the world.

Meanwhile the Institute of Botany of Chinese Academy of Sciences is making living collections of desert plants in the hope of understanding their ecology and finding better ways to propagate and spread them.

In many cases it is not tree cover that will hold back the advancing sand but the layer of small sand tolerant plants that binds the sand in its root system and starts the formation of a protective soil and turf layer. Shading by planted trees could in fact inhibit this process and worsen the problem.

The situation is not helped by the impacts of overgrazing which breaks up the ground turf allowing wind to cut underneath the existing turf and also climate change which produces more climatic extremes of heat, drought as well as rain and wind.

Mixed conifer forests formerly covered most of north east China.

以往, 中国东北大部分地区覆盖着混合针叶林。

"三北"防护林建设

1978 年，中国政府决定建设"三北"防护林体系。"三北"地区系指华北北部、东北西部和西北大部，它东起黑龙江省宾县，西至新疆乌孜别里山口，东西长 4480 公里，南北宽 560～1460 公里，横跨 13 个省的 551 个县，总面积达 40690 万公顷，占中国国土总面积的 42.4%。

历史上"三北"地区是茂密的森林和肥美的草原，千百年来，30 多个民族、1.5 亿人口在这里生息繁衍。由于历代战争的破坏，近代人口的激增，大面积森林和草原遭到破坏。到 70 年代中期，全区平均森林覆盖率仅为 5%，其中荒漠、半荒漠地区仅为 0.96%。"三北"地区社会经济基础薄弱，自然条件复杂多样，十年九旱，灾害频繁，大部分地区年降水量都在 400 毫米 以下，有些地方不足 50 毫米，这里分布着塔克拉马干等 12 个大沙漠、沙地，总面积达 1.33 万公顷。

成了万里风沙线，造成沙进人退的局面。这里有世界上水土流失最严重、最集中区域之一的黄土高原，面积达 3670 万公顷，年流失土壤 10～100 吨/公顷，使中民族的摇篮——黄河每年河床升高 5～10 厘米，造成下游河床高出地面 10 米 以上，成了巨大隐患。

"三北"防护林建设工程用 73 年时间(1978～2050 年)，分三个阶段实施：第一阶段计划用 23 年造林 2177.4 万公顷，建成一批区域性防护林体系；第二阶段用 20 年时间造林 801.7 万公顷，防护林体系基本成形；第三阶段用 30 年时间造林 523.3 公顷，主要是打几个攻坚战和巩固完善提高。到 2050 年，使"三北"地区森林面积增加到 6057 万公顷，森林覆盖率从 70 年代中期的 5% 提高到 15%，林木蓄积量提高 43 亿立方米，生态经济效益累计可达 1.3 万多亿元。

从 1978 年启动到 2004 年，"三北"防护林工程共完成造林 2350 万公顷，成为世界林业建设史上持续时间最长、范围最广的生态建设工程。

在此期间，中国科学院植物研究所在收集活着的沙漠植物，希望通过了解它们的生态关系，找到更好的繁殖与扩散方法。

在许多情况下，真正阻止沙漠前进的不是乔木，而是低矮的沙生植物，它们用自己的根系将流沙固定，开始形成保护性的土壤和草皮层。在乔木的树荫下，实际上不利于沙生植物的生长，从而会加剧荒漠化。

使荒漠化问题更加严重的，是过度放牧和气候变化影响。过度放牧会破坏地面的草皮，风能够刮到现存草皮的底部，而气候变化会导致更多炎热、干旱、大雨和风灾等极端气候。

The government programme to return steep farms to forest or grassland cover, aims to halt the severe erosion and soil loss from such steep fields in areas of high rainfall.

政府把陡峭的农场还原成森林或草原，目的是防止降雨量高的陡峭地区发生严重的土壤侵蚀和水土流失。

Restoring Wetlands

In 2002 The State Forestry Administration, together with 16 other agencies and assisted by the Worldwide Fund for Nature (WWF), published the China National Wetland Conservation Action Plan. The plan included a review of the important wetlands of China, significant species of those wetlands and identified a plan of actions required to better protect and monitor those wetlands.

Among other actions the need to actively restore and rehabilitate wetlands was agreed and from that date there has been a reversal of the former pattern of loss of wetlands through drainage, impoundment and pollution. The urgency in re-establishing lakes in the Yangtze Valley was spurred by the terrible damage from the floods in 1998. It has been decided to restore Dongting lake to its 1949 size of 4,350 km2, an increase of almost 50% from its current summer size. This required breaking of many polders and finding alternative livelihoods for the farmers who have been gradually reclaiming this land. In other areas farmers are being paid to block drains formerly dug to drain wetlands to create more farmland.

A specific wetlands project is conducted since 1999 under funding from the Global Environmental Facility (GEF) to support wetland conservation in general and four important sites specifically—the Sanjiang Plains of north-eastern Heilongjiang, the high altitude peat bogs of the Ruoergai Plateau spanning the Sichuan-Gansu border, Dongting Lake in the Yangtze river basin in Hunan, and the Yancheng coastal marshes and inter-tidal flats of Jiangsu.

Prior to that the Poyang Lake Nature Reserve, which is so important for the Siberian Cranes, was supported under another GEF project managed by the World Bank and WWF had been supporting encouraging wetlands management in China for many years through training programmes and other assistance via its excellent Mai Po Marshes centre in Hong Kong.

6.5 % of the land area of China is classed as wetlands, half of which are natural wetlands. By the end of 2007 China has established 535 wetland reserves totaling 14.5 million ha and including 30 RAMSAR sites of globally important wetlands.

The Rouergai marshes extend across the borders of Sichuan, Gansu and Qinghai provinces and serve as an upper water source for both the Yangtze and Yellow rivers.

若尔盖湿地位于四川、甘肃和青海省的边界，是长江和黄河的源头。

恢复湿地

2002 年，国家林业局连同其他 16 个机构，在世界自然基金会的帮助下，制定了《中国湿地保护行动计划》，对中国重要湿地和重要物种进行了描述，并制定了保护与监测这些湿地的行动计划。

除了采取保护行动外，还达成了一致意见：需要积极恢复和重建湿地。从那时起，因开垦、修筑堤坝和污染导致湿地丧失的格局被逆转了。1988 年的特大洪水造成了巨大的损失，这增强了长江流域湖泊重建的紧迫性。决定将洞庭湖恢复到 1949 年时的面积，即 4,350 平方公里，相当于把目前夏季湖面面积增加 50%。为了实现这个目标，需要进行退田还湖，因而需要为耕作这些土地的农民找到替代生计。在其他地区，政府出资让农民堵塞以前开垦湿地时挖掘的排水沟渠，来恢复湿地。

从 1999 年开始，全球环境基金实施了一个专门的湿地保护项目，并在四个重要地点进行了示范。这四个地点包括：黑龙江东北部的三江平原、四川-甘肃边界地区的若尔盖高海拔泥炭沼泽地、湖南的洞庭湖和江苏盐城沿海沼泽地与滩涂。

在此之前，由于鄱阳湖自然保护区对保护白鹤至关重要，获得了世界银行实施的另一个全球环境基金项目。世界自然基金多年来一直支持中国湿地的管理，并通过香港米埔自然保护区野生生物教育中心提供人员培训项目及其他援助。

中国的湿地占国土面积的 6.5%，其中天然湿地占一半。到 2007 年年底，中国已经建立起 535 个湿地保护区，总面积 1450 万公顷，其中有 30 个湿地保护区已经列入国际重要湿地名录。

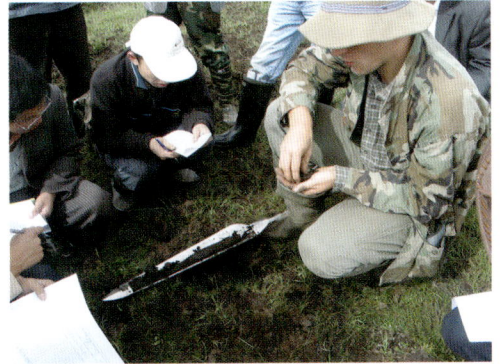

ECBP field project team out investigating Rouergai peatlands. the peat is essential as a huge store of carbon for climate amelioration.

中国－欧盟生物多样性项目的野外项目小组在调查若尔盖泥炭地。泥炭是碳的巨大储存库，是调节气候必不可少的元素。

The great marshes of Rouergai are an important grazing area, important carbon peat store, important water catchment of both Yellow and Yangtze rivers and important for biodiversity.

若尔盖沼泽地既是一个很重要的放牧区，又是重要的泥炭储备库，也是黄河和长江两江的重要集水区，更是生物多样性重要保护地。

Ramsar Programme

The Ramsar Convention or International Convention on Wetlands was established in 1971 and China joined this convention in 1992 and originally nominated 7 sites of international importance. This number has subsequently been raised to 30. These sites have a combined area of 3.43 million hectares and make up 9.4 percent of the country's natural wetland area.

The Ramsar Convention Implementing Office is housed in the State Forestry Administration and serves as a cornerstone of their programme for protecting wetlands in China.

The Eurasian Spoonbill is a water bird that sifts sideways through the water and mud finding small creatures to eat. It breeds in Russia and north China but winters down the east coast of China.

白琵鹭是一种水禽，它们嘴往两侧甩，在水和泥里寻觅小动物吃。它们在俄罗斯和华北繁殖孕育，但在中国东海岸越冬。

拉姆萨湿地公约

《拉姆萨湿地公约》或《国际湿地公约》建立于 1971 年，中国于 1992 年加入该协定，最初提名了 7 个具有国际重要性的湿地，后来增加到 30 个。这些湿地的总面积为 343 万公顷，占全国天然湿地面积的 9.4%。

《拉姆萨湿地公约》执行办公室设在国家林业局，它中国湿地保护工作中发挥着关键的作用。

Wildlife Protection

Apart from the specific programmes to save wildlife species both in nature and in artificial breeding centres, wild animals are given general legal protection through the Wildlife Protection Act. The act provides differing level of protection depending upon which schedule a species is listed and prohibits or regulated the killing, collection, trade or breeding of protected species.

The lists are reviewed and revised periodically on the basis of the global status assessments in the IUCN process of Red Listing of endangered species and on the international agreements on trade made by CITES (see below).

China's Wildlife Protection Act of 1988 updates the old schedules of protected animals and strengthens the penalties that can be applied. There is even a death penalty for killing or trading in the parts of species listed on state category I protection list. In fact this extreme penalty has, to date, only been used for the killing of giant pandas and elephants.

Confiscated chiru skins in the Changtang nature reserve. The soft wool is smuggled to India to make expensive luxury shahtoosh shawls.

羌塘自然保护区收缴的藏羚皮。藏羚羊的柔软绒毛被走私到印度，做成昂贵奢侈的沙图什披肩。

野生动物保护

除了在野外和人工养殖中心实施野生动物保护专项外，还颁布了《野生动物保护法》对野生动物进行普遍的法律保护。《野生动物保护法》依据被列出的物种清单，确定了不同的保护等级，并对受保护的野生动物的捕杀、猎捕、贸易或繁殖进行限制和管理。

在国际自然保护联盟关于濒危物种的全球现状评估的基础上，根据《濒危野生动植物种国际贸易公约》关于国际贸易的协定，定期地对保护野生动物名录进行检查和修改。

1988颁布的中国《野生动物保护法》更新了保护动物名录，加强了对违法者的处罚力度。捕杀和买卖国家一级保护动物甚至要被判处死刑。实际上，直到今天，这种极刑只用于捕杀大熊猫和大象的案例。

CITES

China became a signatory of the Convention on Trade in Endangered Species (CITES) in 1981 and promptly appointed the then Ministry of Forestry to be the management agency for its implementation with Chinese Academy of Sciences serving as the scientific authority.

In line with the obligations of the convention, China added to the list of protected species all those national species not yet protected in China that were listed internationally on the three appendices of CITES. Thus some general groups such as orchids and cycads became protected in China together with black bear, wolf and other species which had formerly been available for hunting and farming.

CITES remains an important global mechanism for documenting and regulating international wildlife trade and China remains a staunch participant and important member in view of the fact that it was formerly such an important trader being a major importer of elephant ivory, rhinoceros horn, other wildlife for food and medicine and also an exporter of saiga antelope horn, tiger parts, and bear gall.

Leaves of ginko tree turn gold in Autumn.

银杏树叶在秋季变为金黄色。

濒危野生动植物种
国际贸易公约

中国于1981年成为《濒危野生动植物种国际贸易公约》的签约国，并立即指定当时的林业部作为公约的管理机构，中国科学院作为科学机构。

根据公约规定的义务，中国把列入《濒危野生动植物国际贸易公约》附录，但尚未得到保护的国内物种确定为保护对象。因而，一些普通物种，例如兰花、苏铁、黑熊、狼以及一些以前允许捕猎的物种也成了受保护对象。

《濒危野生动植物国际贸易公约》是一个对国际野生动物贸易进行记录与调节的重要全球机制。以前，中国是一个大的野生动物进口国，进口象牙、犀牛角、其他野生动物用于食用与药用，中国也是一个羚羊角、虎骨产品和熊胆等的出口国，因此，中国是《濒危野生动植物国际贸易公约》的重要成员国，一直积极参与公约的活动。

Environmental Impact Assessment and Strategic Environmental Assessment

环境影响评估和规划环评

It is recognized that many of China's major developments such as construction of highways, railways, pipelines, airports, canals as well as construction of industrial and residential complexes all have impacts on natural habitat and wild species.

Legislation is now in place under the Ministry of Environmental protection to ensure that such developments do not take place without prior approval of an Environmental Impact Assessment (EIA) that should determine not only if the development will cause unacceptable levels of pollution, noise or disturbance to human communities but also any negative impacts on the natural environment.

It is possible to ask developers to incorporate mitigating measures to limit or counter any such negative impacts.

On an even large scale Strategic Environmental Assessments (SEA) are being conducted for regions, provinces, river basins and even nationally to determine areas where development should be limited for environmental reasons and zoned for different levels of protection.

The China Blueprint Project conducted by The Nature Conservancy together with Ministry of Environmental Protection, reviews the distribution of remaining wild habitata, existing protected areas, distribution of important species and distribution of human threats to identify gaps in the environmental protection system and priority areas for further conservation action.

在中国，人们已经认识到许多重大的开发活动，例如建设高速公路、铁路、管道、机场、隧道、工业与居民建筑物，都会影响到天然栖息地和野物种。

在环境保护部的推动下，关于环评的立法已经开始实施。环评立法要求在此类开发项目上马之前，必须进行并通过环境影响评估。环境影响评估不但要确定发展项目是否污染和干扰人类社区，是否制造噪声，而且还要确定是否会对自然环境带来不利影响。

有可能要求开发者采取缓解措施，减少或消除任何这种负面影响。

规划环评的范围更大，可以是地区、省、流域，甚至国家。规划评估的目的是为了确定是否应为了环境保护而限制发展项目和分区，进行不同程度的保护。

为了寻找环境保护系统的不足之处，确定以后保护行动的重点领域，美国大自然保护协会和环境保护部正在实施中国蓝图项目，对现存天然栖息地、现有保护区、重要物种和人类威胁的分布进行了评论。

ECBP field project team out investigating Rouergai peatlands. the peat is essential as a huge store of carbon for climate amelioration.

中国-欧盟生物多样性项目的野外项目小组在调查若尔盖泥炭地。泥炭是碳的巨大储存库，是调节气候必不可少的元素。

Preserving Agrobiodiversity

The fruits of good practice - disease free, high quality pomelos grown on grafts to original semi-wild fruit trees.

良好做法的成果——无病虫害、高质量的柚子，结在嫁接过的半野生果树上。

China is one of the world's richest nations in terms of its agricultural biological diversity. It is one of the countries of origin of wild rice of which 50,000 local varieties are cultivated and of soya bean which has 20,000 varieties. Buckwheat is another important cereal of Chinese origin. More than 11,000 species of medicinal plants are also known in China and 4,215 species of forage plants. China is the source of wild litchi, kiwi fruits, chestnuts, apples and hundreds of other edible plants and fruits. China is the origin of 30 genera and 2,238 species of ornamental plants and has recorded 1938 varieties of domestic animals.

On 23rd November, 1970, Yu Longping's assistant Li Bihu and Feng Zhishan, a technician from Hainan Nanhong Farm, found a wild rice plant with male sterility, thus name "Yebai (Wild Sterility)". From this plant, Yu Longping created hybrid rice. By 2006, China's hybrid rice had covered over 5.6 billion mu (1 mu equals 1/15 hectare), with rice yield increasing by 520 billion kilograms. In the past years, hybrid rice cultivation has been maintained on about 240 million mu, which is able to feed 70 million people in China. By the end of 2007, hybrid rice had widely spread in more than 20 countries and regions over the world, including Vietnam, Brazil, Bangladesh, Liberia, Guinea and USA, covering an area of over 30 million mu.

Since the survey on wild rices from 1978 to 1982, their habitats have become severely damaged. Wild rice plants in China are divided into 3 categories: *O. rufipogon*, *O. officinalis* and *O. meyeriana*, but the majority of their distribution localities have disappeared and the habitat of the surviving minority are under harsh conditions.

Wild rice plants have gone extinct in Hainan Nanhong Farm where the wild rice plant was found initially.   Wild rice in China is concentrated in Guangdong and Hainan, and 80% of the 1182 localities of *O. officinalis* have disappeared; and only 3%-5% of the habitats of *O. meyeriana* have survived; 12.9% demes of *O. meyeriana* have gone extinct and 83.9 percent remain in moderate and severe external disturbance.

Maliutang Guigang City Guangxi was the world's largest concentration of *O. rufipogon*, but has now disappeared for ever.   Experts estimate that in 20 years, wild rice will go extinct in the wild.

15% of the land area of China is classed as farmland. Another 15% is plantations. These farmlands, fruit and timber plantations, gardens and ponds are home to a vast array of valuable agricultural biological diversity (ABD). These are the species man has long selected from nature for special care and artificial selection. These are the species and varieties upon which we most depend for our daily needs. The variety of these resources is a huge treasure and needs careful guarding.

The history of agriculture in China dates back 7000 years and written records of Chinese Traditional medicine date back to The Book of Songs 2,600 years ago. China is the origin of domestication of horses, yaks, goats, pigs, ducks, pheasants and many other familiar farm animals. China is also the home of horticulture and early records describe the hundreds of varieties of peony and chrysanthemum grown in imperial gardens.

ABD provides not only the foods for 1.3 billion humans but also medicines and industrial products as well as ecological services of climate control, water and soil conservation, nutrient recycling and environmental cleansing.

An increasing urbanization across China includes the creation of green spaces—gardens, parks, golf courses for recreational uses but these too can contain a wide range of valuable biodiversity and perform important ecological services.

Rape fields paint the landscape bright yellow over wide areas in the Sichuan Spring. Rape provides vegetables, seeds for oil and fodder for animals.

四川的春天，油菜花把田野刷成一片鲜黄。油菜一身是宝，可做蔬菜，可榨油，还可用作饲料喂养动物。

保护农业生物多样性

中国是世界农业生物多样性最丰富的国家之一。中国也是野生稻的起源国之一，当地种植的水稻品种达50,000个，大豆品种达20,000个。荞麦是源自中国的另一重要谷物。在中国，已知的药用植物有11,000多种，饲料植物有4,215种。中国是野荔枝、猕猴桃、板栗、苹果和几百种其他食用植物和水果的原生地。中国是30属和2,238种观赏植物的起源地，记录的家畜品种达1938个。

野生稻和杂交稻

1970年11月23日，袁隆平的助手李必湖和海南南红农场技术员冯志珊，在三亚一片野生水稻中，发现了一株雄花败育的野生稻，取名"野败"。就是从这棵野稻上，袁隆平培育出了杂交水稻。

到2006年，我国累计推广种植杂交水稻56亿多亩，增加亩产5200多亿公斤。近年来，全国杂交水稻年种植面积2.4亿亩左右，全中国年增产的稻谷可以养活7000多万人口。

杂交水稻截至2007年底已在全世界包括越南、巴西、孟加拉国、利比里亚、几内亚、美国等二十多个国家和地区推广，种植总面积达到三千多万亩。

1978年至1982年全国野生稻普查以来，野生稻的自然生存环境受到严重破坏。我国拥有普通野生稻、药用野生稻、疣粒野生稻3个野生稻品种的大多数分布点已经消失，尚存的分布点呈残存状态。在最初发现"野败"的海南三亚南红农场，已经见不到野生稻的踪迹。

在我国野生稻资源最丰富的广东和海南，普通野生稻的1182个分布点已消失了80%；药用野生稻保存下来的分布点也只是原来的3%至5%；疣粒野生稻已有12.9%的居群灭绝，83.9%的居群处于中度和重度外界干扰之下。广西贵港市麻柳塘曾经是世界上最大的普通野生稻连片栖地，现在都已不复存在。专家预测，20年后野生稻将在野外消失。

中国有15%的土地面积被划分为农用地，另有15%的土地面积是人工林。这些农用地、果林、用材林、花园和池塘是大量珍贵农业生物多样性的家园。这些物种是人们在长期的生产实践中，从自然界精心挑选出来的，是日常生活所依赖的物种和品种。这些资源的多样性是一大财富，需要好好保护。

中国农业的历史可以追溯到7000年以前，传统中药的文字记录可以追溯到2600年以前的《诗经》。中国是驯养马、牦牛、山羊、猪、鸭、雉类和许多其它家畜的起源地。中国也是园艺之乡，仅皇家花园内种植的有记录的牡丹和菊花品种多达几百个。

农业生物多样性为13亿中国人民不仅提供了食物、药物和工业品，也提供了气候调节、水土保持、养分循环和环境净化等生态服务。

中国不断增长的城市化发展创建了许多绿色空间，包括，花园、公园和高尔夫球场。它们不仅是休闲的场所，也包括广泛的珍贵生物多样性并提供着重要的生态服务。

Education and Awareness

Knowledge and attitudes about wildlife in China remain basic. Most people enjoy seeing animals in zoos, may eat animals or keep animals as pets but have little realization about the role of healthy ecosystems in regulating China's water flow, climate, creating soils or recycling nutrients.

There is a huge need for more environmental training in the school programme and additional need for a wide range of awareness activities to improve the understanding of the general public and also government decision makers to ensure that these matters are given due consideration in the planning and decision making processes in China.

A recent case highlights the growing importance of the voice of the people. In the booming southern city of Xiamen, 10,000 protestors lit a democratic spark by standing up for the environment. And they won. After the Xiamen city government introduced a chemical plant project that was expected to simultaneously generate money (around US$10.45 billion), pollute lakes, and cause cancer local demonstrators sent more than 1 million text protests and 10,000 people held protests around the government offices. Construction was halted. Efforts to make a by-law banning the protests failed and eventually the government decided to abandon the construction of the PX factory. Scholars predict that the win over Xiamen PX will leave a longterm impact.

ECBP Field projects put great emphasis on participation of local communities.

中国－欧盟生物多样性项目的地方项目非常重视当地社区的参与。

教育和意识

在中国，关于野生动物的知识和对野生动物的认识还很基本。多数人们喜欢到动物园观看动物，可能会吃野味，或将动物当宠物养，但对健康生态系统在调节水流和气候，生成土壤和循环养分方面的作用了解甚少。

非常有必要在中小学教育中增加更多环境教育的内容，同时也要开展丰富多彩的宣传教育活动，来提高公众以及政府决策者的环保意识，以确保在规划和决策过程中，对环境保护给予充分的考虑。

公众舆论与厦门 PX 项目

厦门 PX 化工项目是 2006 年厦门市引进的外资总额达 108 亿元人民币的腾龙芒烃（厦门）有限公司的一个化工项目。项目选址于厦门市海沧区，项目投产后每年可为厦门市增加 800 亿元人民币的工业产值。但在今年 3 月全国两会上，中科院院士赵玉芬等 105 名全国政协委员联名提交了一份"关于厦门海沧 PX 项目迁址建议的提案"。提案指出，离居民区仅 1.5 公里的 PX 项目存在泄漏或爆炸隐患，厦门百万人口面临危险，必须紧急叫停项目并迁址。

ECBP in action - active debate among students of Beijing universities.

中国－欧盟生物多样性项目在行动——组织北京的大学生积极辩论生物多样性。

Ruddy shelduck landing on lake. This is a hardy species of northern and western China.

赤麻鸭降落在湖上。它不怕寒冷，分布在中国北部和西部。

Agencies Involved in Biodiversity Conservation

The National Steering Committee for the Convention of Biological Diversity consists of 24 ministries and other agencies. The box below lists the major players among these and their main areas of responsibility.

- Ministry of Environmental protection
  (ecological conservation and general coordination)
- State Forestry Administration
  (nature reserves, reforestation, wildlife, 'Grain for Green'programme, wetlands conservation, etc.)
- Ministry of Water resources
  (soil conservation, water resources, river basin organization)
- Ministry of Agriculture
  (grassland management, agro-biodiversity, fisheries and aquatic animals)
- National Development and Reform Commission
  (rural economic development, West China Developemnt, Planning)
- Ministry of Construction
  (urban construction, communications and also World heritage issues)
- State Oceanography Administration
  (marine fisheries and marine conservation areas)
- Chinese Academy of Sciences
  (scientific authority, research, collections, documentation, monitoring)

In addition to these national bodies, a number of international organizations and GAAs are actively involved in biodiversity conservation programmes in China.

- European Union-biodiversity conservation and integrated river basin management
- Asian Development Bank-ecosystem management in Baiyangdian, Sanjiang, Ningxia
  and degrading of dry ecosystems
- UNDP/GEF-wetland conservation and sustained use
- World Bank -ecological restoration in Loess plateau, reforestation
- AusAID -reforestation in Qinghai
- GTZ-reforestation and agro-biodiversity conservation
- DFID, WI, and others -range of conservation projects and activities

Cock Hume's pheasant sunning itself.

黑颈长尾雉雄鸟在晒太阳。

Korsac fox is a small canid of the barren north of the country.

沙狐是小型犬科动物，栖息在华北开阔的半沙漠地带。

The World Conservation Union (IUCN)

IUCN is huge international union of state, agency, NGO and individual members; based in Gland, Switzerland and with regional offices and some national offices around the world. They have released a series of Best Practices guidelines on many aspects of environmental conservaton. They have recently launched their Countdown 2010 programme to bolster efforts to reach the millennium goal objectives set for that year by the United Nations.

Worldwide Fund for Nature (WWF)

WWF were the first international conservation NGO to start working in China following the initial 1980 cooperation to study and save Giant Pandas in Wolong. Today WWF has grown and broadened its interest with a wide range of programmes all across China. This covers conservation education, training, field projects, seeking alternative livelihood for inhabitants of biorich areas and working on policy issues relating to wetlands and other issues.

The Nature Conservancy (TNC)

TNC is a large US-based conservation organization, whose mission is to preserve the plants, animals and natural communities that represent the diversity of life on Earth by protecting the lands and waters they need to survive. TNC use strategic, science-based planning processes to identify the highest-priority places—landscapes and seascapes that, if conserved, promise to ensure biodiversity over the long term. In China TNC have been working to preserve biodiversity in NW Yunnan with the cooperation of local villagers.

Conservation International (CI)

CI is another fast-growing US-based conservationeco-protection organization with a mission to conserve the Earth's living heritage, our global biodiversity, and to demonstrate that human societies are able to live harmoniously with nature. The organization is involved in field surveys, integrated conservation projects and conservation awareness.

The Wildlife Conservation Society (WCS)

WCS saves wildlife and wild lands through careful science, international conservation, education. These activities change individual attitudes toward nature and help people imagine wildlife and humans living in sustainable interaction on both a local and a global scale. In China, WCS is located within the Chinese Academy of Sciences. They work in some of the remotest parts of China on the Qinghai-Tibetan Plateau and in NE China on the Russian border (Amur Tiger project).

Wetlands International (WI)

Wetlands International is dedicated to wetland conservation and sustainable management and aims to ensure that wetlands and water resources are conserved and managed for their full range of values and services, benefiting biodiversity and human well-being. Wetlands International-China (WI-C) was formed in 1996 is focused on assessment and monitoring of wetlands and wetland taxa to ensure their conservation and sustainable management in China and North Asia.

Fauna and Flora International (FFI)

FFI, located in Cambridge, UK, is acting to conserve threatened species and ecosystems world-wide, choosing solutions that are sustainable, based on sound science and take account of human needs. FFI publishes the respected conservation journal 'Oryx'. FFI work to save entire ecosystem by forming partnership with whoever needs to be involved, from governments and NGOs to industry and local communities. FFI has undertaken projects in China since 1999.

Ink cap fungi are tasty when young.

幼嫩的真菌味道极好。

参与生物多样性
保护的机构

中国实施《生物多样性公约》协调组由24个部委组成。下框中列出了其中主要的参与者及各自的主要职责。

- 环境保护部(生态保护，全面协调)
- 国家林业局(自然保护区、造林、野生动物、退耕还林工程、湿地保护等)
- 水利部 (水土保护、水资源、流域组织)
- 农业部(草地管理,农业生物多样性,渔业和水生动物)
- 国家发展与改革委员会(农村经济发展,西部大开发,规划)
- 建设部(城市建设,通讯和世界遗产问题)
- 国家海洋局(海洋渔业和海洋保护区)
- 中国科学院(科学权威、研究、收集、建立文献、监测)

除这些国家机构外，许多国际组织和GAAs (政府援助机构) 也积极参与了中国的生物多样性保护项目。

- 欧盟——生物多样性保护和综合流域管理
- 亚洲开发银行——白洋淀、三江和宁夏的生态系统管理，干旱生态系统退化
- 联合国开发署 / 世界环境基金——湿地保护和可持续利用
- 世界银行——黄土高原的生态恢复，造林
- 世界自然基金会——森林、淡水、长江物种保护
- 大自然保护协会——计划项目、自然资本、示范项目
- 澳大利亚国际 / 海外发展署——在青海造林
- 德国技术合作公司——造林和农业生物多样性保护
- 英国国际发展部,世界保护联盟,湿地国际,保护国际,世界保护战略、野生动
  植物保护国际及其他一系列保护项目与活动

The male of Hippolimnas butterfly is strikingly coloured, the female mimics distasteful milkweeds.

Hippolimnas 雄蝴蝶具惊艳的色彩，而雌性则模仿味道难吃的蚨蝶。

Great-spotted woodpecker works up tree stem.

大斑啄木鸟在啄虫。

Wild poppies grow at high altitudes on the mountains.

高山上的野罂粟花。

世界自然保护联盟（IUCN）

世界自然保护联盟是巨大的全球性自然保护联盟，有国家，政府机构，非政府组织和个人参与；总部设在瑞士格兰德，在世界各地设有国家办事处和地方办事处。该联盟发布了一系列环保方面的最佳实践准则。最近他们推出了2010年倒计时方案，目的在于督促各方加强努力，以实现2000年由联合国确定的2010年目标。

世界自然基金会（WWF）

WWF在中国的工作始于1980年在卧龙自然保护区的大熊猫及其栖息地的保护，是第一个受中国政府邀请来华开展保护工作的国际非政府组织。今天，世界自然基金会在中国不断壮大，保护兴趣不断发展，现在全国各地都有项目。这些项目包括环保教育，培训，野外项目，为野生资源丰富地区的居民寻求替代生计，并在湿地和其他保护问题的相关政策立法方面开展工作。

大自然保护协会（TNC）

大自然保护协会是总部设在美国的一个大环保组织，协会的使命是：通过保护代表地球生物多样性的动物、植物和自然群落赖以生存的陆地和水域，来实现对这些动物、植物和自然群落的保护。大自然保护协会利用合作性战略，用科学原理来指导保护规划进程，以确定最优先保护的陆地和海洋景观，以确保生物多样性的长期持续性。大自然保护协会在中国一直致力于与当地村民合作，维护滇西北的生物多样性。

保护国际（CI）

保护国际是另一个快速增长的保护组织，总部设在美国，其宗旨为保护地球上尚存的自然遗产和全球的生物多样性，并证明人类社会和自然是能够和谐共处的。该组织参与中国环保工作的实地调查，综合养护和提高保护意识等项目。

野生动物保护协会（WCS）

野生动物保护协会通过科学，国际保护以及教育来保护野生动物和野生地。他们的活动无论在地方还是全球范围内都改变了个人对自然世界的态度，帮助人们想象野生动物和人类的可持续的相互作用。在中国，野生动物保护协会设在中科院。他们的工作在中国最偏远的青藏高原和中国东北的中俄边境（东北虎项目）。

湿地国际（美国）

国际湿地组织致力于湿地保护和可持续管理，目的是确保湿地和水资源的养护和管理，以充分发挥湿地的价值观和服务，有利于生物多样性和人类福祉。湿地国际中国办事处成立于1996年，重点是评估和监测湿地和湿地分类，以确保中国和北亚湿地的保护和可持续管理

野生动植物保护国际（FFI）

野生动植物保护国际，总部设在英国剑桥，致力于在科学的基础上，充分考虑人类的需求，选择可持续性的解决方法，以保护全球的濒危物种和生态系统。FFI出版颇受尊重的保护期刊'Oryx'。FFI通过与地方机构：从政府机构或非政府机构到工企或地方社区，建立伙伴关系来实现对整个生态系统的保护。自1999年以来FFI就在中国开展了项目。

Pearls of dew cling to the fragile spiders web.

纤纤蜘蛛网上挂着晶莹的露珠。

Laws and Regulations

There is no comprehensive law to govern the preservation and management of biodiversity in China. A mix of laws is currently required to control the use of lands, forests, wildlife and nature reserves. The details below are only a few of the majorinstruments. There are many other relevant laws and ordinances.

The Environmental Protection Law 1989 is formulated to protect and improve people's environment and the ecological environment, preventing and controlling pollution and other public hazards, safeguarding human health and facilitating the development of socialist modernization.

Article 17 calls for the protection of regions representing various types of natural ecological systems, regions with a natural distribution of rare and endangered wild animals and plants, regions where major sources of water are conserved, geological structures of major scientific and cultural value, famous regions where karst caves and fossil deposits are distributed, traces of glaciers, volcanoes and hot springs, traces of human history, and ancient and precious trees. Damage to the above is strictly forbidden.

Within the scenic spots or historic sites, nature reserves and other zones that need special protection, no industrial production installations that cause environmental pollution shall be built.

The law calls for the undertaking of environmental impact assessments for any developments liable to cause pollution or damage to the environment. Such assessments should stipulate suitable preventive and curative measures to be taken.

The Law on the Protection of Wildlife was enacted in 1988. It establishes rules for wildlife management and protection. This law provides for penalties against violations, including the death penalty for killing some first grade protected species. The lists of protected species thereon are periodically updated. According to Article 16 of the law, catching or hunting wildlife listed under the "first" grade of protection requires a special license, which is granted only if the killings are "necessary for scientific research, domestication and breeding, exhibition or other special purposes". Unfortunately some provincial forestry offices take this to include trophy hunting and much debate follows. The guiding policy of the law is to "protect, develop, and rationally utilize wildlife resources," with an emphasis on breeding, domestication, and development rather than protection. A recent review of the law calls for revisions drop the "protection for human use" objective, expand protection scope, introduce anticruelty provisions, and nationalize protection responsibilities.

The Forestry Act of 1998 declares the objectives of the forestry sector including plantations, nature reserves, wildlife conservation and sustained management of natural forests. The law targets 30% of forest cover for the whole land area. Many of the natural forests are severely degraded, even those areas inside nature reserves and forest parks, and in need of some sort of enrichment management or fallow protection. This has become easier since the logging bans of the late 1990s and the closing of many timber farms.

1992 Regulations on Nature Reserves define and prescribe the establishment of nature reserves with three zones—strictly protected core area, relatively unmodified surrounding buffer zone and man-modified experimental zone. The regulations follow the MAB design but that was established for studying the relationships between Man and his Environment not for nature protection per se. The Chinese regulations are confusing in that in most international schemes, the buffer zone is a highly modified external zone and not a well protected internal zone. The regulations also provide little flexibility for design or management options

The white stork is a rare bird of northeast China. It has been helped through building of artificial nests. This one is eating a small bird.

白鹳是中国东北的一种珍稀鸟类，为白鹳搭建人工巢对它的生存有很大的帮助。这只白鹳正在吃一只小鸟。

The black-eared toad is a large poisonous species.

黑框蟾蜍就是我们常说的癞蛤蟆，皮肤分泌出的白色乳液有毒，经过加工即是中药"蟾酥"。

Chinese water deer watches alert and ready to spring away in flight.

獐警觉地张望，一有风吹草动就会逃之夭夭。

so that although in reality many nature reserves in China face rather high levels of human use and modification, there are no suitable zones and only one category for nature reserves.

The need to overhaul these regulations, establish more permissible zones and a range of protected area categories based on the IUCN classification system has been repeatedly proposed but still never agreed upon.

Wetlands Protection regulations define wetlands and identifies the State Forestry Administration as the key agency for their protection. However, the law does not give overall management authority to Forestry so that many contradictions exist with Ministry of Agriculture handling issuance of fishing and reed cutting permits, Department of water resources controlling sluice levels and water take off, Ministry of Environmental Protection responsible for monitoring pollution and discharge into waterways and Ministry of Transportation organizing shipping lanes and communications.

相关法规

中国目前还没有保护和管理生物多样性的综合法律，必须援用在土地利用、林业、野生生物和自然保护方面的多种专门法。以下介绍的是其中的几个主要法律。另外还有其他一些相关的法律法规。

1989年颁发的《环境保护法》其目的是保护和改善生活环境与生态环境，防治污染及其他公害，保障人体健康，促进社会主义现代化建设的发展。

第17条要求保护各种类型的自然生态系统区域，珍稀、濒危的野生动物自然分布区域，重要的水源涵养区域，具有重大科学文化价值的地质构造、著名的溶洞和化石分布区、冰川、火山、温泉等自然遗迹，以及人文遗迹、古树名木，应当采取措施加以保护，严禁破坏。

在景区、历史遗址、自然保护区和其他需要特别保护的区域内，不得装建会导致环境污染的工业生产设施。

《环境保护法》要求对有可能造成污染或环境破坏的发展项目进行环境影响评估并规定防治措施。

《野生动物保护法》于1988年颁布，翌年实施，它确立了野生动物管理和保护的规则。违法都将受到法律的惩罚，其中捕杀某些一级保护动物者可被判处死刑。保护动物名录将定期修订。第16条禁止猎捕或杀害国家重点保护野生动物，因科学研究、驯养繁育、展览或其他特殊情况需要捕捉、捕捞国家一级保护野生动物的，必须申请特许猎捕证。不幸的是，一些省林业官员利用这一条，为来中国狩猎的外国猎人颁发许可证，并因此引发了许多争论。《野生动物保护法》规定，国家对野生动物实行加强资源保护，积极驯养繁殖，合理开发利用的方针，鼓励开展野生动物科学研究。在最近一次对《野生动物保护法》的审议中，有人提出将其中的"为人类利用而保护"的目的删除，将保护的范围扩大，引入了禁止虐待动物的条款，并将保护责任国民化。

1998年的森林法宣布了林业行业的目标，包括造林、自然保护区、野生动物保护和天然林的可持续经营。法案中明确提出要将森林覆盖率提高到国土面积的30%。许多天然林，甚至包括一些自然保护区和森林公园内的天然林，已严重退化，需要进行补植或封山育林。自从上世纪九十年代末实施"天保工程"以来，关停了很多森工企业，森林保护变得容易多了。

1992 年的自然保护区管理条例对自然保护区进行了定义和描述，它包括三个区域：核心区，缓冲区和实验区，其中缓冲区是人类活动相对较少的区域，而实验区则是人类活动较多的区域。条例套用了人与生物圈计划（MAB）对自然保护区的设计，而没有认识到，人与生物圈计划目的是为了研究人与环境之间的关系，而非为了自然保护。因此，在国际惯例中，缓冲区是位于保护区之外，其中可以允许进行频繁的人为活动，而中国的条例却把缓冲区定义为保护区内部应该受到严格保护的一个区域。另外，该条例把所有的保护区域都按照自然保护区来对待，因此在区划和管理措施上没有选择的余地，而实际上中国的许多自然保护区内都存在着严重的人为活动和干扰。

尽管曾有组织机构提议对这些条例进行全面检查修订，并根据世界自然保护联盟（IUCN）的分类系统划分区域和建立各类保护区，但是各方却一直未真正达成共识。

湿地保护条例对湿地进行了定义，并确定国家林业局作为湿地保护的主管部门构。然而，法律并未将全部的管理权力授予国家林业局，因而与颁发捕鱼和采割芦苇许可证的农业部，控制蓄排水量的水利部，负责监测污染和控制向水道排放废水的环境资源部，以及组织船运和交通的交通部之间，都存在着矛盾。

The dipper rests on a snowbound stream bank.

河乌在积雪的溪边停留小憩。

Public responsibility

The government develops the broad strokes for action—the national policy and strategy for biodiversity conservation. The government develops appropriate laws and under those laws different agency issue more detailed regulations governing management, utilization, protection of biological resources. The government even launches and funds some major programmes for conserving biodiversity and improving ecological services.

But ultimately the rules and policies are only words and ideals. What counts is action on the ground and action on the ground is undertaken by real people.

The day to day actors are the farmers, and poachers, the traders and foresters, the police and school teachers, the tourists and general public, NGOs and village communities. What really determines whether China will save its biodiversity, preserve natural habitats, rare species and beautiful natural heritage is the level of public responsibility that is achieved.

If the public ignore the issue and imagine it is the government's sole responsibility to take the steps necessary for biodiversity conservation, then China will lose many wonderful species, will lose valuable ecosystem services will have less development options in the future and face greater numbers of natural catastrophes.

We need the public to work hand in hand with the government. They should have enough awareness to understand the issues, support government policy and aspirations and play their respective roles in achieving a green China.

The public should want a cleaner, greener environment. They must demand and respect green spaces. Avoid pollution and littering. Become less wasteful with respect to water and energy resources; become more discerning in their choices about foods, furs, commodities they buy and items they wear.

They should learn to appreciate and take pride in their green spaces, natural recreation areas, forest parks, heritage parks, nature reserves.

Lycaenid butterflies patiently ensure their species survival.

lycaenid 蝴蝶，耐心地确保其物种的生存。

Red-crowned crane flying over reed beds.

丹顶鹤飞越芦苇地。

White-cheeked starling attacks autumn persimmon crop.

灰椋鸟在秋季饱餐柿子。

It is everyone's job. The banker should be more open to loaning money for environmentally sound enterprises, the magistrate should realize the importance of infringements of environmental regulations, the town planner realize the need for sustainable parks, the golf course superintendent realize his golf course can serve as a nature reserve as well as a sports ground. Broadcasters should see the need to promote greater news and awareness on environmental issues, schools should deliver sounder grounding to children to understand the importance of biodiversity. Businesses should be more supportive of and more involved in environmental protection. Tourism operators should ensure that fair share of income flows into the communities and locations that serve as tourism destinations.

China faces a huge awareness task and educators, media, NGOs government and international agencies must all play their parts in delivering this greater awareness.

公共责任

政府为生物多样性保护行动采取了重大举措: 制定了关于生物多样性保护的国家政策和方案。政府制定了适当的法律, 根据这些法律, 不同的机构又制定了关于生物资源管理、利用和保护详细条例。政府甚至启动和资助了一些关于保护生物多样性和生态环境建设的重大工程。

但是, 最终这些规定和政策只是文字与想法而已, 关键在于要有具体的人把它们付诸实施。

日常的行动者是农民、偷猎者、商人、林业工作人员、执法者与学校教师、游客和公众、非政府组织和地方社区。中国能否拯救生物多样性, 保护天然栖息地、稀有物种和美丽的自然遗产, 取决于公众的认识程度和采取行动的决心。

如果公众忽视这个问题, 把保护生物多样性的责任完全推给政府, 那么中国将丧失许多极好的物种, 将失去宝贵的生态系统服务, 将来的发展机会将更少, 还要面对更多的自然灾害。

我们需要公众要与政府密切合作, 他们应该有基本的环保意识, 对这些问题有清楚的认识, 支持政府政策和号召, 并在建设绿色中国中发挥自己应有的作用。

公众需要一个清洁的、绿色的环境。他们必须追求和尊敬绿色空间。避免污染和乱扔杂物。减少对水资源与能源的浪费, 在衣食住行中要增强环保意识, 作出明智的选择。

绿化中国, 人人有责。银行家应该更开明地为环境绩效良好的企业贷款; 政府官员应该意识到违反环境规章的严重性; 城乡规划者应意识到可持续公园的必要性; 高尔夫球场的管理者应意识到高尔夫球场既是一个运动场也是一个自然保护区。传媒应该播放更多有关环境问题的新闻和教育片; 学校应该提供更多机会让孩子们理解生物多样性的重要性; 商人应该更积极地支持和参与环境保护; 旅游业经营者应该保证与社区和旅游点公平地共享旅游收入。

在中国, 提高和培养保护意识仍然是一项紧迫的重大任务, 教育者、媒体、非政府组织政府和国际机构都必须发挥各自的作用, 为增强公众的环保意识做出贡献。

EU official hands out prizes to the winning team of the universities students biodiversity debate organised by ECBP.

中国·欧盟生物多样性项目的官员给大学生生物多样性辩论优胜队颁奖。该活动由中国·欧盟生物多样性项目主办。

Representative of Ministry of Environmental Protection in television interview.

中国环境部代表接受电视采访。

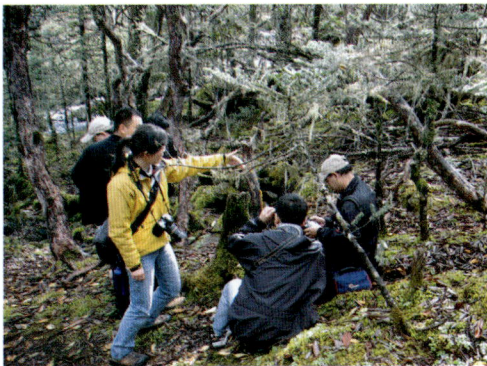

Field surveys in Laojun Mountain of northwest Yunnan as part of the ECBP programme.

中国·欧盟生物多样性项目在滇西北老君山实地调查。

EU-China Biodiversity Programme

One of China's largest and boldest conservation initiatives is the cooperative programme known as the EU-China Biodiversity Programme (ECBP) between the European Union (EU), United Nations Development Programme (UNDP), the Chinese Ministry of Commerce (MOFCOM) and The Chinese Ministry of Environmental Protection (MEP). The programme combines policy dialogue and development, institutional strengthening and awareness raising with a set of field projects to improve the effectiveness of biodiversity conservation at the provincial and local level. EU contributes € 30 million, of which € 21 million is earmarked for field projects, which also require a minimum of 50% matching funds from partnerships of national and international organizations. Eighteen field projects have been short listed through a professional and transparent tender procedure and most of these have been selected and are currently under implementation. The field projects will develop and disseminate best practice models for sustainable management of key ecosystems and integrate biodiversity into sectoral planning at provincial and local levels.

The Programme was signed in June 2005 and officially launched on International Biodiversity Day, May 22, 2006. It is currently scheduled to end in March 2010.

The Results and Resources Framework of the project lays out five results:

1. Programme activities are regularly planned, monitored and evaluated

2. Strengthened national policy and institutional framework for biodiversity conservation

3. Biodiversity mainstreamed into relevant policies and legislation, including SEA and EIA

4. Increased awareness of decision-makers and general public of biodiversity and its importance

5. Increased capacity for biodiversity conservation at provincial and local level through 18 partnership field projects in central, southern and western China.

The ECBP programme is now in full swing and the field projects up and running. We can only hope these are successful and develop new approaches for implementation of conservation actions at local levels. Lessons learned from the field can be channeled back into central policy thinking.

Meanwhile the value of China's Green Gold can be better evaluated, better appreciated and better safeguarded as a solid foundation for China's continuing impressive development.

欧盟－中国生物多样性项目

欧盟-中国生物多样性项目是中国规模最大的保护行动之一，由欧盟、联合国开发署、中国商务部和中国环境保护部合作实施。项目融合了政策对话和制定，机构强化，意识增强和一批地方示范项目，旨在提高各省与地方各级政府保护生物多样性的效力。欧盟援助3000万欧元，其中2100万欧元专门用于地方示范项目，要求合作国家与国际组织至少提供50%的配套经费。通过专业透明的投标程序，最终确定了19个地方示范项目。多数项目已经获得支持进行整个项目书的完善，到2008年1月，有14个示范项目已经开始实施。通过实施地方示范项目，将开发和推广重点生态系统可持续管理的最佳实践模型，并将把生物多样性纳入省和地方的行业规划。

项目于2005年6月签定，于2006年5月22日的国际生物多样性日启动，计划于2010年3月结束。

项目的结果与资源架构中设定了五个方面的结果：

1. 有规律地对项目活动进行规划、监测和评估
2. 改善关于生物多样性保护的国家政策和强化机构
3. 将生物多样性纳入相关的政策和立法，包括策略环境评估和环境影响评估
4. 提高决策者和公众对生物多样性及其重要性的认识
5. 通过在中国中部、南部和西部合作实施18个地方示范项目，提高省和地方各级政府保护生物多样性的能力。

中国-欧盟生物多样性项目工作已全面展开，我们期望这个项目圆满成功，开发和推广可在地方一级实施的环境保护方案，而地方项目总结出的经验应纳入中央决策规划。

同时，中国绿色宝藏应得到更好的评价，更好的欣赏，更好的保护，使之继续为中国飞速发展奠定坚实的基础。

EU representative meeting Yunnan official at biodiversity workshop.

生物多样性研讨会上，欧盟代表与云南官员会谈。

Recommended bibliography
推荐参考书目

The ripening fruit of the Hawthorn bush is used as a tonic and traditional medicine.

快熟了的山楂。该果实常被用作补品和传统药物。

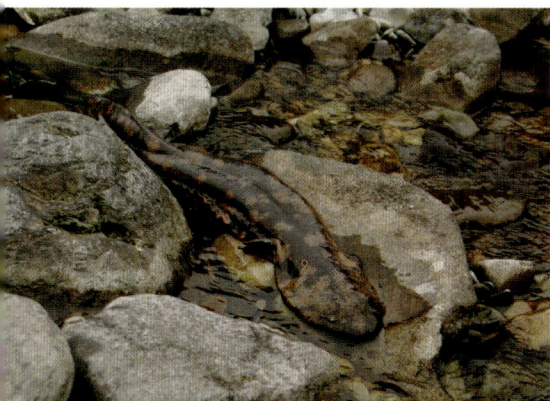

Giant salamanders live in the smaller branches of the Yangtze River. Several species are all threatened by collection as food, by pollution and fragmentation of habitat.

中国大鲵生活在长江的小支流中。有几个种类都因被大量打捞作为食品，或因污染及栖息地被破碎而受到威胁。

MacKinnon, J & K Phillipps. 2000. *A Field Guide to the Birds of China*, Oxford University Press
MacKinnon, J & K Phillipps. 2000. 中国鸟类野外手册. 牛津大学出版社

Schaller, G.B. 1993. The Last Panda. University of Chicago Press, Chicago.
Schaller, G.B. 1993. 最后的熊猫. 芝加哥大学出版社

Schaller, G.B. 1977. The Mountain Monarchs – Wild Sheep and Goats of the Himalaya. University of Chicago Press, Chicago.

Schaller, G.B. 1998. Wildlife of the Tibetan Steppe. University of Chicago Press, Chicago.
Schaller, G.B. 1998. 青藏高原上的生灵. 芝加哥大学出版社

Xie Yan (ed). 2008. Biodiversity and Ecological Security in China. Hebei Science and Technology Press, Beijing.
解焱 编. 2008. 生物入侵与中国生态安全. 河北科学技术出版社

Smith, A. T. and Xie Yan (eds). 2008. *A Guide to the Mammals of China*. Princeton University Press, Princeton and Oxford.
Smith, A. T. 解焱等编. 2008. 中国哺乳动物指南. 普林斯顿大学出版社

Xie Yan, W. Sung & P. Schei (eds.) 2004. China's Protected Areas. Tsinghua University Press, Beijing.
解焱 汪松 P. Schei 等编. 2004. 中国的保护地. 清华大学出版社

MacKinnon, J. & N. Hicks. 1996. Wild China. New Holland, London.

Harris, R. 2008. Wildlife Conservation in China. M.E. Sharpe Inc., Arizona.
Harris, R. 2008. 中国的野生生物保护. 亚利桑那州 M.E. Sharpe 出版社.

Zhao Songqiao. 1986. Physical Geography of China. Science Press, Beijing and John Wiley & Sons Inc, New York.
赵松桥. 1986. 中国自然地理. 北京科学出版社及纽约 John Wiley & Sons 出版社.

Wilson, E.H. 1913. A Naturalist in Western China. Methuen & Co Ltd, London.

Wang Chi-wu. 1961. The Forests of China with a Survey of Grassland and Desert Vegetation. Harvard University Press, Cambridge, Massachusetts.

Acknowledgements:

Thanks are due to the European Union and other key stakeholders of the EU-China Biodiversity Programme for their funding and implementation of the programme and to the individual projects and staff of the programme for their help in logistics, flow of information, stories and photos. Agreco provided backstopping support and advice and the EU delegation staff in Beijing encouragement and technical back-up.

Map credits: All maps were provided by Cai Bofeng

Photo credits: All photos are provided by ECBP, VAC unless otherwise stated.

Xie Yan: cover; 4 lower left; 5 lower; 6 upper right; 7 upper left, upper centre; 21 upper; 27 upper, lower; 39; 42 lower; 78; 79; 94; 102; 121; 128; 129 upper; 133; 134; 140; 162; 163 upper; 175; 202; 204; 211; 217 upper; 224; 225 lower; 226; 227; 228 lower; 229 upper, lower; 244; 246; 259 upper; 274 upper; 277; 300 upper.

George Schaller: 80; 223; 225 upper.

Xi Zhinong: 7 lower right; 8; 37 upper, lower; 60; 68; 108; 109; 110; 116; 124; 125; 126; 127 upper; 129 lower; 138; 141; 142; 143; 150; 151; 184; 191; 206; 258; 270; 276; 303 lower.

Liu Jin: 55 upper, lower; 235; 236 lower; 237; 238 upper, lower.

Andrew Laurie: 7 upper right; 294; 299 upper.

Nigel Hicks: 29 lower.

Wang Pu: 283 lower.

Conservation International: 187; 290.

Wetlands International: 77; 201; 282; 283 upper; 287.

Zhao Chao FFI: 111.

WCS: 285.

GTZ: 222.

TNC: 89; 209.

ADB: 180.

ECBP, COSU: 23 upper, lower; 54; 200; 279.

鸣谢：

衷心感谢欧盟的资助以及中国 - 欧盟生物多样性项目（以下简称"中欧项目"）其它利益相关方对项目的资助和实施；非常感谢地方示范项目与中欧项目员工的协助，提供的相关信息、故事与照片。同时也感谢 Agreco 的支持和宝贵意见；最后感谢北京欧盟驻华代表团工作人员的鼓励以及技术支持。

地　图：由蔡博峰提供

照　片：除有特殊说明，所有照片由中欧项目宣传教育子项目提供。

解　焱：封面；4 页左下；5 页下；6 页右上；7 页左上，中上；21 页上；27 页；39 页；42 页下；78 页；79 页；94 页；102 页；121 页；128 页；129 页上；133 页；134 页；140 页；162 页；163 页上；175 页；202 页；204 页；211 页；217 页上；224 页；225 页下；226 页；227 页；228 页下；229 页；244 页；246 页；259 页上；274 页上；277；300 页上。

乔治·夏勒：80 页；223 页；225 页上。

奚志农：7 页右下；8 页；37 页；60 页；68 页；108 页；109 页；110 页；116 页；124 页；125 页；126 页；127 页上；129 页下；138 页；141 页；142 页；143 页；150 页；151 页；184 页；191 页；206 页；258 页；270 页；276；303 页下。

刘津：55 页上，下；235 页；236 下；237 页；238 页。

Andrew Laurie：7 页右上；294 页；299 页上。

Nigel Hicks：29 页下。

王　普：283 页下。

保护国际：187 页；290 页。

湿地国际：77 页；201 页；282；283 页上；287 页。

赵　超，野生动植物保护国际：111 页。

国际野生生物保护学会：285 页。

德国技术合作公司：222 页。

大自然保护协会：89 页；209 页。

亚洲发展银行：180 页。

中欧项目国家办公室支持机构：23 页上，下；54 页；200 页；279 页。

图书在版编目（CIP）数据

绿色中国／马敬能，王海滨编著.-北京：中国文联出版社，
2008.12
ISBN978－7-5059-6180-7

Ⅰ.绿… Ⅱ.①马…②王… Ⅲ.生态环境-环境保护-中国 Ⅳ.X321.2

中国版本图书馆CIP数据核字(2008)第187265号

书　　名	绿色中国
编　　著	马敬能　王海滨
出　　版	中国文联出版社
发　　行	中国文联出版社 发行部 （010-65389150）
地　　址	北京农展馆南里10号(100125)
经　　销	全国新华书店
责任编辑	刘　旭
责任校对	李睿芝
责任印制	焉松杰　刘　旭
印　　刷	北京华联印刷有限公司
开　　本	889×1194　1/12
印　　张	25.5
插　　页	2页
版　　次	2008年12月第1版第1次印刷
书　　号	ISBN978－7-5059-6180-7
定　　价	198.00元

您若想详细了解我社的出版物
请登陆我们出版社的网站http://www.cflacp.com